ENERGY SCIENCE, ENGINEERING AND TECHNOLOGY

NATURAL GAS SYSTEMS

ENERGY SCIENCE, ENGINEERING AND TECHNOLOGY

Additional books in this series can be found on Nova's website under the Series tab.

Additional E-books in this series can be found on Nova's website under the E-book tab.

ENERGY SCIENCE, ENGINEERING AND TECHNOLOGY

NATURAL GAS SYSTEMS

RAFIQ ISLAM
EDITOR

Nova Science Publishers, Inc.
New York

For permission to use material from this book please contact us:
Telephone 631-231-7269; Fax 631-231-8175
Web Site: http://www.novapublishers.com

Library of Congress Cataloging-in-Publication Data

Natural gas systems / editor, Rafiq Islam.
p. cm.
Includes index.
ISBN 978-1-61324-158-5 (softcover)
1. Gas wells. 2. Natural gas pipelines. 3. Gas manufacture and works. I. Islam, Rafiqul, 1959-
TN880.N297 2011
338.2'7285--dc22

2011008471

Published by Nova Science Publishers, Inc. † New York

CONTENTS

PREFACE

This new book presents and discusses research in the study of natural gas systems. Topics discussed include flammability and individual risk assessment for natural gas pipelines; guidelines for developing gas fields associated with edge-water drive; fuzzy estimation and stabilization in gas life wells based on a new stability map and human health risks assessment due to natural gas pipelines explosions.

Chapter 1 - The objective of this work is to investigate the effects of key reservoir and aquifer parameters on gas recovery factor for finite and infinite aquifers. The first part deals with performance calculations of cumulative water influx and cumulative gas production at successive time steps. The effects of aquifer size to reservoir size ratio, rate of pressure depletion at the gas/water contact, initial pressure at the gas/water contact, initial gas in place, angle of water encroachment, and volumetric sweep efficiency on the cumulative water influx and cumulative gas production are investigated. In the second part, a new analytical approach is proposed to determine abandonment pressure and gas recovery factor for the case of infinite aquifers. The third part caters for performance calculations of gas recovery in the case of finite aquifers. For this purpose, another new analytical approach is suggested for estimating the rate of pressure depletion at the gas/water contact.

It is concluded that the reservoir and aquifer parameters considered in this study can significantly affect, to a varying degree, the amounts of water influx and gas production. For infinite aquifers, higher rates of pressure depletion at the gas/water contact are found to yield higher gas recovery factors, less number of lbm-moles of remaining gas in the reservoir, and lower abandonment pressures. For finite aquifers, higher ratios of aquifer size to reservoir size are shown to yield higher abandonment pressures and larger number of lbm-moles of remaining gas in the reservoir at abandonment conditions. The gas recovery factor, however, is found insensitive to the ratio of aquifer size to reservoir size. The results of this work can be of great value to reservoir and production engineers dealing with this type of gas reservoirs and should provide them with the necessary guidelines for planning their production strategies.

Chapter 2 - System dynamics stability has been used to construct stability maps for gas-lifted oil wells. A fuzzy TS model is constructed based on proposed stability maps and Fuzzy Kalman Filter concept is used to estimate immeasurable state variables in noisy environment of well operation. Then, Fuzzy Linear Quadratic Gaussian regulator is constructed by combination of fuzzy estimator and fuzzy TS controller to stabilize the highly oscillatory well operations. The quadratic stability of the overall closed-loop system is evaluated in terms of Lyapunov with a less conservative criterion. Simulations are performed using a transient multi-phase flow simulator and variable input command signals are introduced to the system. Satisfactory results are obtained in terms of estimation and stabilization.

Chapter 3 - Several papers describe as sour gases should be treated but the energy consumption in the process is a problem is being encountered. The energy consumption can be affected for the type of amine used in the absorption process (the reboiler can need more or less steam). Presently at Petrotrin, the results shown in this article indicated that the MDEA shows a lower reboiler duty for regeneration, such that the use of Methyldiethanolamine, MDEA is economically feasible and as such, it is recommended that the existing amine Diethanolamine or other amine mixtures system be replaced by the use of Methyldiethanolamine, MDEA.

Chapter 4 - Natural gas and oil are mainly supplied and transmitted through pipelines. The safety and risk factors for transporting natural gas through pipelines is an important issue. This paper develops a comprehensive model for the individual risk assessment for natural gas pipelines. Presently available models related to pipeline risk assessment are also examined and their shortcomings identified. To overcome these limitations, a new concept of individual risk is introduced. It combines the flammability limit with existing individual risk for an accidental scenario. The new model determines the major accidental area within a locality surrounded by pipelines. Finally, the proposed model is validated using field data. This innovative model applies to any natural gas pipeline risk assessment scenario.

Chapter 5 - Natural gas is transported mainly by pipelines throughout the world. Therefore it is necessary to assess and manage the resulting risks regarding human health issues due to gas toxicity and flammability. It is possible to assess the risk of irreversible damage to a human being for any accidental scenario by introducing specific vulnerability functions. Events such as flash fire, vapor cloud explosion, and fire can be understood by the maximum predicted amount of vapor within the flammability limits for the entire history of its dispersion. Another danger to human health lies in the flammability of natural gas transportation systems. A human health risk assessment study in the event of such an accident has been carried out in this paper. In this study, a 1 to 20% accidental rate is considered for assessing individual risk due to flammability. A newly developed flammability risk management model is used in the present study. The research shows that the individual risk due to the flammability of natural gas is not more than an 18 percent human health hazard. The findings of this study will be helpful to improve health hazard risk management and remediation.

Versions of these chapters were also published in *Advances in Sustainable Petroleum Engineering,* Volume 1, Numbers 1-3, published by Nova Science Publishers, Inc. They were submitted for appropriate modifications in an effort to encourage wider dissemination of research.

In: Natural Gas Systems
Editor: Rafiq Islam
ISBN: 978-1-61324-158-5

Chapter 1

GUIDELINES FOR DEVELOPING GAS FIELDS ASSOCIATED WITH EDGE-WATER DRIVE

Hazim H. Al-Attar*
Chemical & Petroleum Engineering Department,
United Arab Emirates University, Al-Ain, UAE

ABSTRACT

The objective of this work is to investigate the effects of key reservoir and aquifer parameters on gas recovery factor for finite and infinite aquifers. The first part deals with performance calculations of cumulative water influx and cumulative gas production at successive time steps. The effects of aquifer size to reservoir size ratio, rate of pressure depletion at the gas/water contact, initial pressure at the gas/water contact, initial gas in place, angle of water encroachment, and volumetric sweep efficiency on the cumulative water influx and cumulative gas production are investigated. In the second part, a new analytical approach is proposed to determine abandonment pressure and gas recovery factor for the case of infinite aquifers. The third part caters for performance calculations of gas recovery in the case of finite aquifers. For this purpose, another new analytical approach is suggested for estimating the rate of pressure depletion at the gas/water contact.

It is concluded that the reservoir and aquifer parameters considered in this study can significantly affect, to a varying degree, the amounts of water influx and gas production. For infinite aquifers, higher rates of pressure

* hazim.alattar@uaeu.ac.ae

depletion at the gas/water contact are found to yield higher gas recovery factors, less number of lbm-moles of remaining gas in the reservoir, and lower abandonment pressures. For finite aquifers, higher ratios of aquifer size to reservoir size are shown to yield higher abandonment pressures and larger number of lbm-moles of remaining gas in the reservoir at abandonment conditions. The gas recovery factor, however, is found insensitive to the ratio of aquifer size to reservoir size. The results of this work can be of great value to reservoir and production engineers dealing with this type of gas reservoirs and should provide them with the necessary guidelines for planning their production strategies.

Keywords: gas reservoirs, edge-water, drive mechanism, reservoir performance, material balance, sensitivity analysis, aquifer size, sweep efficiency, water influx, abandonment pressure, depletion rate.

NOMENCLATURE

B_{gabn}	gas FVF at abandonment pressure p_{abn}, ft^3/STB [res m^3/std m^3]
B_g	gas FVF at pressure p, ft^3/STB [res m^3/std m^3]
B_{gi}	gas FVF at pressure p_i, ft^3/STB [res m^3/std m^3]
B_w	water FVF, RB/STB [res m^3/stock-tank m^3]
c_f	pore compressibility, psi^{-1} [Pa^{-1}]
c_t	total compressibility, psi^{-1} [Pa^{-1}]
c_w	water compressibility, psi^{-1} [Pa^{-1}]
e	rate of reservoir/aquifer boundary pressure depletion, psia/day [Pa/day]
E_g	underground gas expansion, ft^3/scf [res m^3/std m^3]
f	(encroachment angle)° / 360
F	total gas and water production, ft^3 [res m^3]
G	original gas in place, scf [std m^3]
G_p	cumulative gas produced, scf [std m^3]
G_{pabn}	cumulative gas produced at abandonment pressure, scf [std m^3]
h	net thickness, ft [m]
p	pressure, psia [Pa]
p_{abn}	abandonment pressure, psia [Pa]
p_i	initial reservoir pressure, psia [Pa]
p_n	reservoir pressure at time t_n,psia [Pa]

r_a	radius of aquifer, ft [m]
R_D	dimensionless radius = r_a/r_g
S_{gr}	residual gas saturation, fraction less than one
S_{wc}	connate water saturation, fraction less than one
S_{wi}	initial water saturation in the gas reservoir, fraction less than one
t	time variable, hrs
t_D	dimensionless time (see Eq. 1)
t_{Dj}	dimensionless time at time step j (see Eq. 2)
T_D	dimensionless time at time step n (see Eq. 2)
T	reservoir temperature, °F [°C]
W_e	cumulative water influx, RB [m^3]
$W_D(t_D)$	dimensionless cumulative water influx function giving the dimensionless influx per unit pressure drop imposed at the reservoir/aquifer boundary from $t = 0$ up to T_D
W_p	cumulative water produced, RB [m^3]
Z	gas deviation factor
Z_{abn}	gas deviation factor at abandonment pressure
Z_i	gas deviation factor at initial reservoir pressure

Greek Letters

α	volumetric sweep efficiency, per cent
Δ	difference
ε	rate of reservoir/aquifer boundary pressure depletion, psia/day [Pa/day]
ϕ	porosity, fraction less than one
γ	gas specific gravity (air = 1.0), dimensionless
μ	viscosity, cp [Pa.s]
Θ	angle of water encroachment, degrees [rad]

Subscripts

a	aquifer
D	dimensionless
e	influx
$emin$	minimum influx
f	formation

g	gas
gi	gas at initial condition
gr	gas at residual condition
gabn	gas at abandonment condition
i	initial condition
j	index of loops
n	number of time steps
p	produced
wc	connate water

SI Metric Conversion Factors

cp x 10^{-3} = Pa.s
degrees x 1.745 329x10^{-2} = rad
ft x 3.048 x10^{-1} = m
ft^3 x 2.831 685 x10^{-2} = m^3
°F (°F - 32)/1.8 = °C
psi x 6.894 757 = kPa
psi^{-1} x 1.450 377 x10^{-1} = kPa^{-1}

1. INTRODUCTION

Predicting the depletion of waterdrive gas reservoirs have been investigated by many authors. Agarwal *et al.* (1965) used a material balance model to study the effect of water influx on gas recovery. They concluded that gas recovery depends on production rate, residual gas saturation, aquifer strength, aquifer permeability, and the volumetric sweep efficiency of the encroaching water zone. Bruns *et al.* (1965) studied the effect of water influx on the *p/z* versus cumulative gas production (G_p) curves. They concluded that it is dangerous to extrapolate the *p/z* charts on a straight line without considering the possibility of water influx. Geffen *et al.* (1952) conducted an experimental study of residual gas saturation under waterdrive. They concluded that residual gas saturation under waterdrive varies from 15 to 50% pore space, depending on the type of sand. Knapp *et al.* (1968) developed a two-phase, two-dimensional model to predict gas recovery from aquifer storage fields. The model was used to study the effects of heterogeneity, aquifer strength, and gas production rates. From their results, they concluded that gas recovery is a function of gas production rate, aquifer strength, and heterogeneity. Their conclusions agree with those of Agarwal *et al.* (1965) in terms of gas production rate and aquifer strength. Shagroni (1977) studied the effect of formation compressibility and edge water on gas field performance. He concluded that it is incorrect to extrapolate the early part of the p/z vs. cumulative gas production curves as a straight line to $p/z = 0.0$, to estimate the initial gas inplace without considering the possibility of water influx and the effect of formation compressibility, and that the sensitivity of the performance curve (p/z vs. G_p) to reservoir compressibility increases as the initial reservoir pressure increases. Pepperdine (1978) used a mathematical model to study the performance of the Devonian gas fields in northern British Columbia. He concluded that to achieve maximum gas recovery, the depletion process should be increased as much as possible by production practices, and that the important factor in the low efficiency of gas recovery was water influx rather than coning phenomenon in the portion of the Clarke Lake field that was modeled. Al-Hashim and Bass Jr. (1988) predicted the depletion performance of partial waterdrive gas reservoirs to study the effect of aquifer size, gas production rate, and initial reservoir pressure on the rate at which the gas-water-contact advances and on gas recovery. Based on their results and for constant reservoir permeability of 300 md, they concluded that regardless of the size of the reservoir, when the ratio of aquifer radius to gas reservoir radius, $r_a/r_g > 2.0$, the pressure in the unsteady-state water influx equation has to be corrected to the original gas-water-contact. They also concluded that gas recovery appeared to be sensitive to initial reservoir pressure and the aquifer size (when $r_a/r_g > 2.0$), and as r_a/r_g and the initial reservoir pressure increase, gas recovery decreases. Their results indicated that gas recovery appears to be sensitive to gas production rate when $r_a/r_g > 3$.

The objective of this study is to investigate the effects of key reservoir and aquifer parameters that have not been thoroughly investigated before, on gas recovery factor for finite and infinite aquifers. The outcomes of this study should provide the practicing engineers with the necessary guidelines mostly needed in the development of gas reservoirs associated with edge-water drive.

2. Theoretical Background

The Van Everdingen-Hurst Model; "The Aquifer Fitting Model"

Van Everdingen and Hurst (1949) proposed solutions to the dimensionless diffusivity equation, and are known as the Constant Terminal Pressure Solution (CTP) and the Constant Terminal Rate Solution (CTR). Engineers prefer the application of the CTP solution in the subject of gas reservoir performance under the effect of water influx, because they are more interested in predicting rates of water influx in terms of a certain pressure drop at the gas/aquifer boundary. The final form of the CTP solution is written as:

$$W_e = U\,\Delta P\,W_D(t_D) \qquad (1)$$

where, U is aquifer constant for radial geometry ($= 1.119 f\phi h c_t r_g^2$) in bbl/psia, W_e is cumulative water influx due to a pressure drop Δp (psia) imposed at the reservoir radius r_g, at time $t = 0$, in bbls, $W_D(t_D)$ is dimensionless cumulative water influx function giving the dimensionless influx per unit pressure drop imposed at the reservoir-aquifer boundary at $t = 0$, f is (encroachment angle)°/360°, which is used for aquifers which subtend angles of less than 360° at the center of the reservoir-aquifer system, ϕ is aquifer porosity fraction, c_t is total aquifer compressibility in psia^{-1}, t_D is dimensionless time (= $2.309\ kt / \phi\mu c_t r_g^2$) and μ is water viscosity in cp.

The dimensionless water influx $W_D(t_D)$ is presented in tabular form or as a set of polynomial expressions giving W_D as a function of t_D for a range of ratios of the aquifer to reservoir radius $R_D = r_a/r_g$, for radial aquifers. In this work the polynomial approach proposed by Klins *et al.* (1988) is used and found much easier to deal with than the lookup tables or charts that may sometimes require interpolations. Polynomial equations are available for finite and infinite aquifers with absolute errors less than 0.03% and 0.02%, respectively.

When applying the van Everdingen-Hurst *aquifer fitting model* in history matching, it is necessary to extend the theory to calculate the cumulative water influx corresponding to a continuous pressure decline at the reservoir-aquifer boundary. In order to perform such calculations it is conventional to divide the continuous decline into a series of discrete pressure steps. For the pressure drop between each step, Δp, the corresponding water influx can be calculated using Eq (1). Superposition of the separate influxes, with respect to time, will give the cumulative water influx, and as follows:

$$W_{e_n}(T) = U\sum_{j=0}^{n-1}\Delta p_j W_D(T_D - t_{Dj}) \qquad (2)$$

This model is then combined with the material balance equation and the resulting equation is used to solve for the single unknown which is the average reservoir pressure at any time step n, (p_n), within the pressure decline history.

If it is felt confident that the aquifer fitting model is satisfactory in matching the history, then the next step is to use it in predicting the future reservoir performance. The aim here is usually to determine how the reservoir pressure will decline for a given gas offtake rate. Knowledge of this decline will assist in calculating the recovery factor, consistent with production engineering and economic constraints. The basic equations are the reservoir material balance and the water influx equation. These can be solved simultaneously, by an iteration process, to calculate the reservoir pressure.

Gas Field Volumetric Material Balance

a) Appropriateness in Application

Whether material balance can be applied to a hydrocarbon accumulation as a whole depends upon how rapidly any pressure disturbance is equilibrated in the reservoir so that it may be treated as zero dimensional. This, in turn, is dependent on the magnitude of the hydraulic diffusivity constant, $k/\phi\mu c$; the larger the value of this parametric group, the more rapidly is pressure equilibrium achieved. It can be shown that, in spite of the high gas compressibility, its extremely low viscosity dominates in making the diffusivity constant many times greater than for oil which enhances the prospect for meaningful application of material balance, even in tight gas reservoirs.

b) Havlena and Odeh (1963) Interpretation

Neglecting water expansion and pore compaction, the material balance equation for gas reservoirs subjected to water influx can be expressed as:

$$F/E_g = G + W_e B_w/E_g \quad (3)$$

where, F ($= G_p B_g + W_p B_w$) is total gas and water production in rcf, E_g ($= B_g - B_{gi}$) is underground gas expansion in rcf/scf, G is initial gas in place in scf, G_p is cumulative gas production in scf, W_e is cumulative water influx in rcf, W_p is cumulative water production in rcf, B_{gi} is gas formation volume factor at initial reservoir pressure in rcf/scf, B_g is gas formation volume factor at current reservoir pressure in rcf/scf, and B_w is water formation volume factor in rbbl/stb, usually set equal one.

Using the production, pressure and PVT data, the left-hand side of this expression should be plotted as a function of the cumulative gas production, G_p. This is simply for display purposes to inspect its variation during depletion.

Plotting F/E_g versus production time or pressure decline, Δp, can be equally illustrative. If the reservoir is of the volumetric depletion type, $W_e = 0$, then the values of F/E_g evaluated , say, at six monthly intervals, should plot as a straight line parallel to the abscissa, whose ordinate value is the GIIP. Alternatively, if the reservoir is affected by natural water influx then the plot of F/E_g will usually produce a concave downward shaped arc whose exact form is dependent upon the aquifer size and strength and the gas offtake rate. Backward extrapolation of the F/E_g trend to the ordinate should nevertheless provide an estimate of the GIIP ($W_e \approx 0$) but the plot can be highly non-linear in this region yielding a rather uncertain result. The main advantage in the F/E_g versus G_p plot, however, is that it is much more sensitive than other methods in establishing whether the reservoir is being influenced by natural water influx or not.

c) p/Z-Interpretation Technique

This is by far the most popular method of applying gas material balance. Neglecting water expansion and pore compaction, the equation is formulated at standard conditions (scf) as:

$$p/Z = p_i/Z_i\{[1 - (G_p/G)] \,/\, [1 - (W_eB_wE_i/G)]\} \qquad (4)$$

The term $W_eB_wE_i/G$ represents the fraction of the hydrocarbon pore volume invaded by water and consequently, the greater the influx the higher the pressure for a given offtake of gas. In the event that there is no influx and the reservoir is of the volumetric depletion type, then the equation may be reduced to the form:

$$p/Z = p_i/Z_i[1 - (G_p/G)] \qquad (5)$$

Equation (5) is a simple linear relationship between p/Z and the fractional gas recovery. It gives rise to the popular field technique of plotting the reservoir averaged values of p/Z , in which the pressures are referred to some common datum level, as a function of the cumulative gas production G_p. If the reservoir is of the volumetric depletion type, then the plot must necessarily be linear, and its extrapolation to the abscissa ($p/Z = 0$) enables the effective GIIP to be determined as $G_p = G$.

Alternatively, if there is natural water influx from an adjoining aquifer, the p/Z plot is, in principle, non-linear. The technique may seem fairly straightforward but this is where the potential danger lies in application of the p/Z plot: deciding what is and what is not a straight line. In great many cases the plot for a water drive field will *appear* to be linear until a very advanced stage of depletion when, in fact, it is not. Consequently, it is suggested that this plot is made with an enlarged *p/Z* scale where the only linear portion of the plot occurs very early in the lifetime of the field, before the water influx is significant and extrapolation of this early trend will give a more reliable value of the GIIP, although it is still likely to be too large.

3. METHOD OF CALCULATION

The calculations were divided into three parts and as follows:

Part I: Calculations of cumulative water influx W_e and cumulative gas production G_p at successive time steps are made for different ratios of reservoir size /aquifer size, pressure depletion rates, initial reservoir pressures, initial volumes of gas in place, and angles of water encroachment. These calculations are accomplished, as described above, by combining two fundamental mathematical models:

A. The Van Everdingen and Hurst aquifer fitting model (1949) (constant terminal pressure solution of the diffusivity equation).
B. Dake (1978) expressed the material balance equation which includes water expansion and pore compaction as
C.

$$G_p = \{G\left[(B_g - B_{g_i}) + \frac{B_{g_i}}{1 - S_{wc}}(c_w S_{wi} + c_f)\Delta P\right] + 5.615W_e\} / B_g \tag{6}$$

where c_w and c_f are water and pore compressibility, psia^{-1}, respectively.

To simplify the calculations, it is assumed that the cumulative water production, W_p, is negligible and the water formation volume factor, B_w, is equal 1.0.

The gas deviation factor, Z, was estimated with the Hall-Yarborough equation (1996) as follows.

$$Z = \frac{0.06125 P_{pr} x \exp[-1.2(1-x)^2]}{y} \tag{7}$$

To solve for Z, however, another empirical equation is used to solve for y using the MS Excel solver:

$$-0.06125P_{pr}x\exp[-1.2(1-x)^2]+\frac{y+y^2+y^3-y^4}{(1-y)^3}$$
$$-(14.76x-9.76x^2+4.58x^3)y^2+(90.7x-24.2x^2+42.4x^3)y^{(2.18+2.82x)}=0$$

Part II: Calculations of gas recovery factors for infinite aquifers; $ra/rg \geq 10$
The following equations are proposed to run the calculations of this part.

$$W_{e\min}=(Original\ pore\ volume)\alpha(1-S_{gr}-S_{wc}) \tag{8}$$

where α is the volumetric sweep efficiency, dimensionless.
Dividing both sides by initial volume of gas in place (G):

$$\frac{W_{e_{\min}}}{G}=\frac{B_{gi}}{1-S_{wc}}\alpha(1-S_{gr}-S_{wc}) \tag{9}$$

The cumulative gas produced at abandonment pressure is calculated as follows;

$$(G_p)_{abn}=Initial\ gas\ in\ place-[Trapped\ \mathrm{Re}sidual\ Gas+Bypassed\ Gas] \tag{10}$$

or

$$(G_p)_{abn}=G-\left[\alpha GB_{gi}(\frac{S_{gr}}{1-S_{wc}})+(1-\alpha)GB_{gi}\right](B_g)^{-1}{}_{abn} \tag{11}$$

The recovery factor is therefore:

$$\frac{(G_p)_{abn}}{G}=F=1-\left[\frac{W_{e\min}S_{gr}}{(1-S_{gr}-S_{wc})GB_{gabn}}+(1-\alpha)\frac{B_{gi}}{B_{gabn}}\right] \tag{12}$$

Agarwal (2004) presented an empirical equation to calculate the residual gas saturation for limestone formations as follows

$$S_{gr} = \frac{A_1(100\phi) + A_2(\log k) + A_3(100S_{gi}) + A_4}{100} \tag{13}$$

where: $A1$ = -0.53482234, $A2$ = 3.3555165, $A3$ = 0.15458573, $A4$ = 14.403977

The value of S_{gr} thus obtained is assumed constant in all calculations.

The amount of remaining gas at abandonment conditions in the reservoir, n, in lbm-mole is calculated by applying the real-gas law as follows;

$$n = \frac{P_{abn}\pi r_g^{\,2} h\phi\alpha S_{gr}}{Z_{abn}(T + 460)R} \tag{14}$$

In this work, a new graphical procedure is proposed to determine the abandonment pressure for this case, as shown in Figure 1. Using the results of part I, plots of We/G versus P are prepared for three different values of pressure depletion rates, e or ε. The value of We_{min}/G then is calculated with Eq (9) and used in the above figure to estimate P_{abn}.

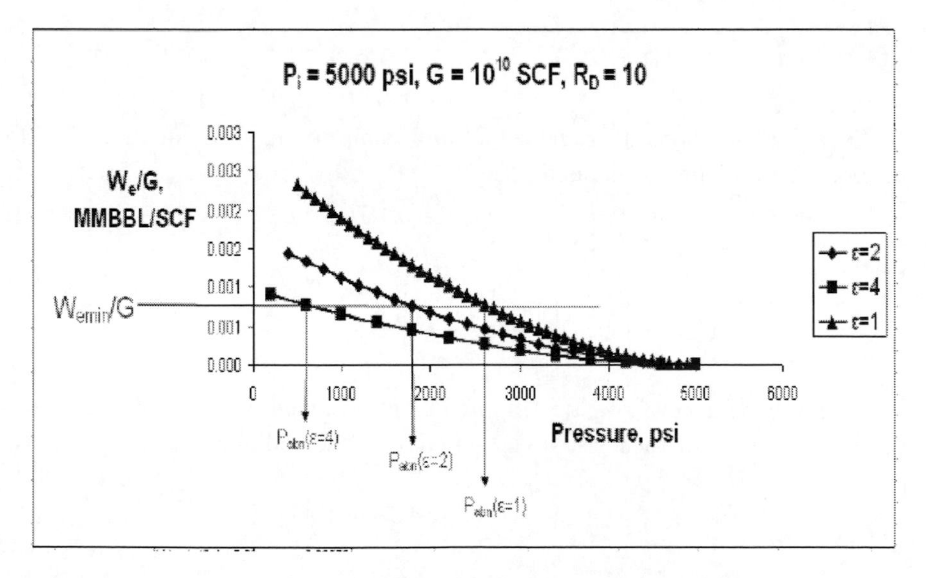

Figure 1. New graphical technique: Determination of p_{abn}-infinite aquifers.

The recovery factor is calculated with Eq (12) and the amount of gas remaining in the reservoir is calculated with Eq (14). Values of recovery factor, F, amount of remaining gas, n, and abandonment pressure, P_{abn}, thus obtained are plotted versus pressure depletion rates, ε. This procedure is applied to investigate the effects of initial reservoir pressure and volumetric sweep efficiency on F, n, and P_{abn}, respectively.

Part III: Calculations of gas recovery factors for finite aquifers; $ra/rg < 10$. The following equations are proposed for the calculations of this part.

$$W_{e\min} = \pi\phi h c_t (r_a^2 - r_g^2)(P_i - P_{abn})\frac{\theta}{360} \tag{15}$$

Since,

$$G = \left[\pi\phi h r_g^2 (1 - S_{wc})\frac{\theta}{360}\right] / B_{gi} \tag{16}$$

Then,

$$\frac{W_{e\min}}{G} = \frac{(c_f + c_w)\left[\frac{r_a^2}{r_g^2} - 1\right](P_i - P_{abn})B_{gi}}{(1 - S_{wc})} \tag{17}$$

Equating equations (9) and (17) and simplifying, an equation for the abandonment pressure is obtained.

$$P_{abn} = P_i - \frac{\alpha(1 - S_{gr} - S_{wc})}{(c_f + c_w)\left[\frac{r_a^2}{r_g^2} - 1\right]} \tag{18}$$

In this part, a new graphical approach is designed to estimate the rate of pressure depletion for each case investigated, as presented in Figure 2.

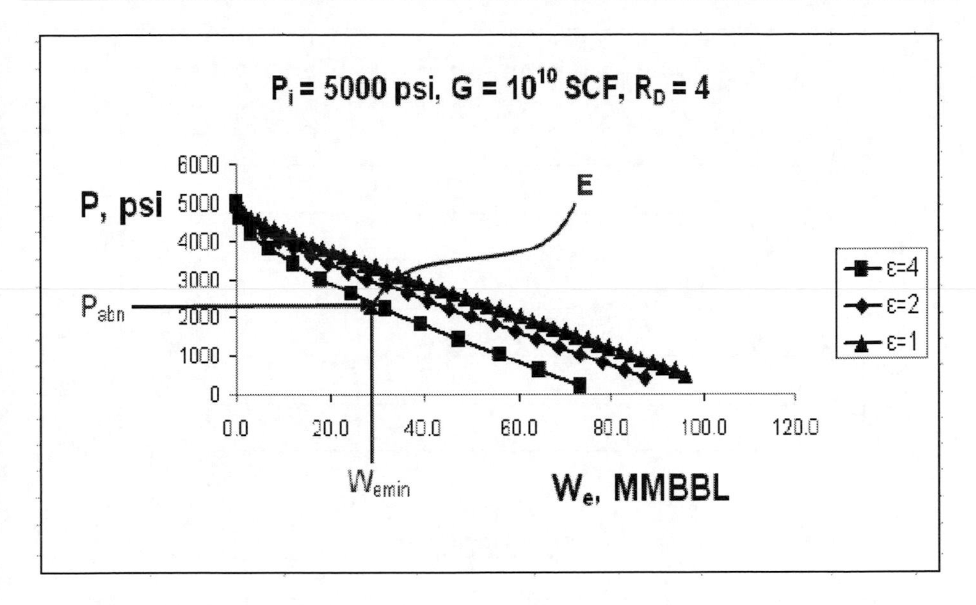

Figure 2. New graphical technique: Determination of ε-finite aquifers.

The values of *Wemin*, *Pabn*, *n*, and *F* are calculated with Eqs. (15, 18, 14 and 12), respectively.

4. Results and Discussion

Due to space limitation, only selected figures are presented in this paper.

Part I: A. Effects of reservoir and aquifer parameters on cumulative water influx (*We*). A base case is selected as a reference for comparisons; $\varepsilon = 2$ psia/day, $p_i = 5{,}000$ psia, $G = 10^{11}$ scf, and $\theta = 45°$.

A.1) Effect of aquifer size to reservoir size ratio (R_D) on *We* ($\varepsilon = 1$, 2, and 4 psi/day, respectively): The results are shown in Figures 3-5. Regardless of R_D, higher rates of pressure depletion reduce the amount of water influx, as the later will have less time available to encroach into the gas reservoir and catch up with the fast moving gas towards the producers. For $R_D \leq 2$ (small aquifer), the effect of the aquifer on gas reservoir performance becomes increasingly insignificant and may even be neglected. For a given value of ε and boundary pressure, larger aquifers ($R_D > 2$) show higher amount of water influx, but this effect is diminished as ε is increased to 4 pisa/day. This implies that *We* appears to be insensitive to R_D as the gas offtake rates become increasingly high.

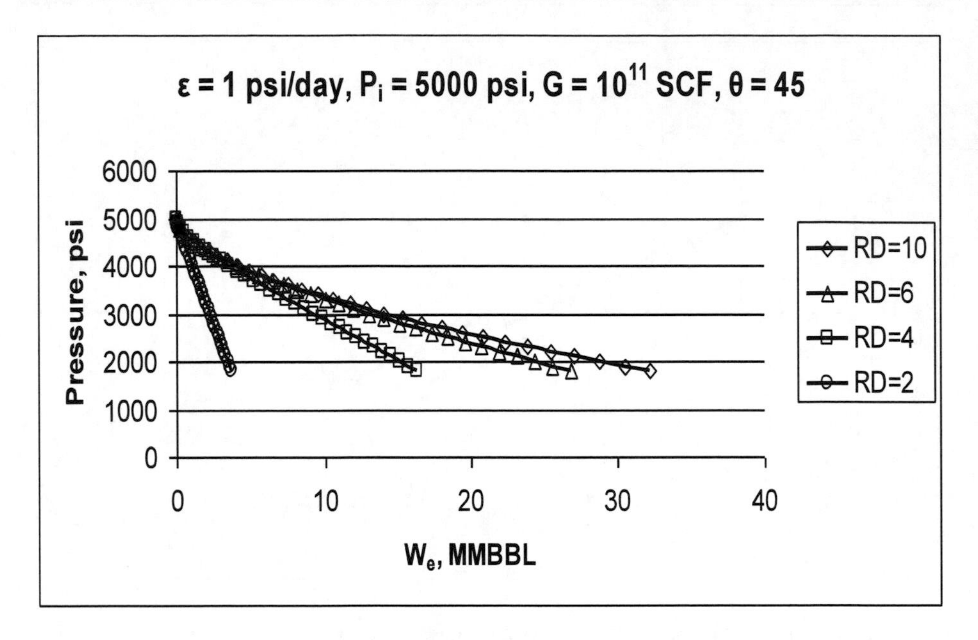

Figure 3. Effect of R_D on We; ε = 1 psi/day.

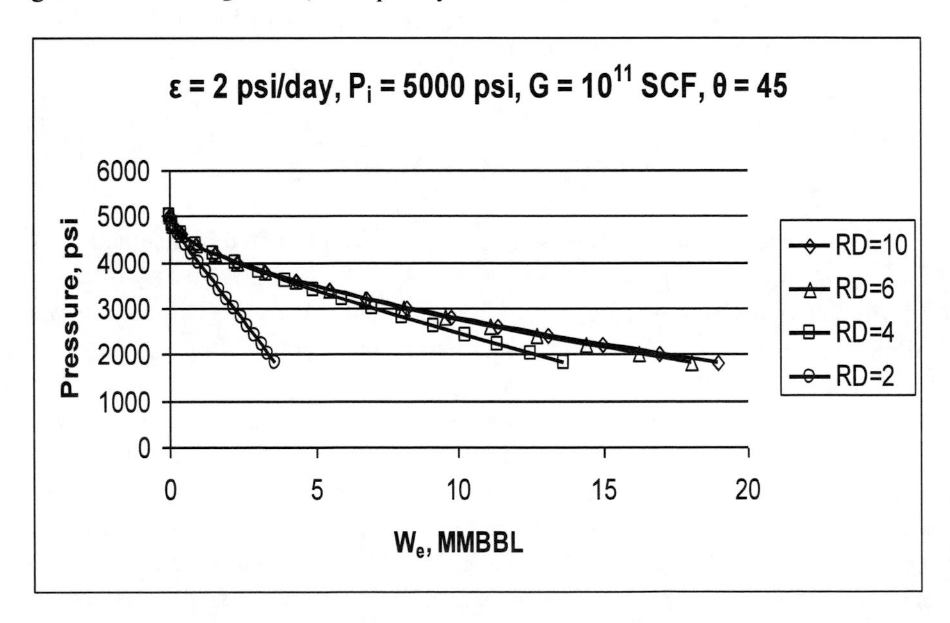

Figure 4. Effect of R_D on We; base case, ε = 2 psi/day.

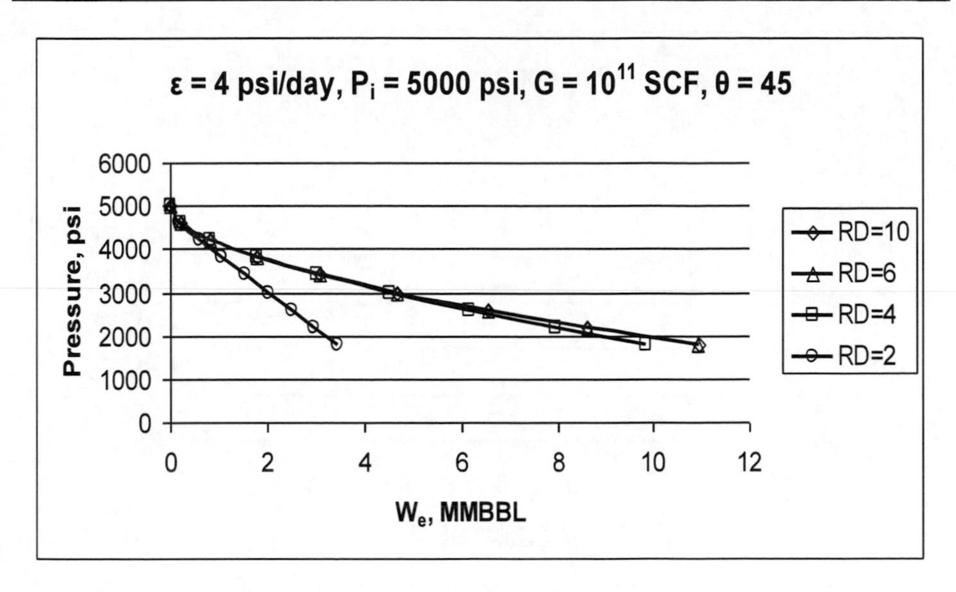

Figure 5. Effect of R_D on We; $\varepsilon = 4$ psi/day.

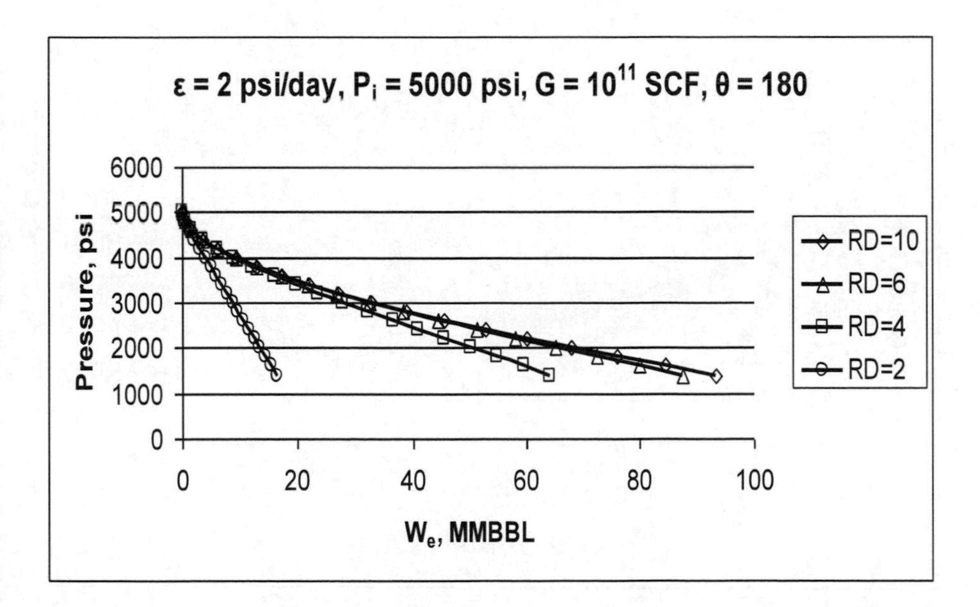

Figure 6. Effect of R_D on We; $\theta = 180°$.

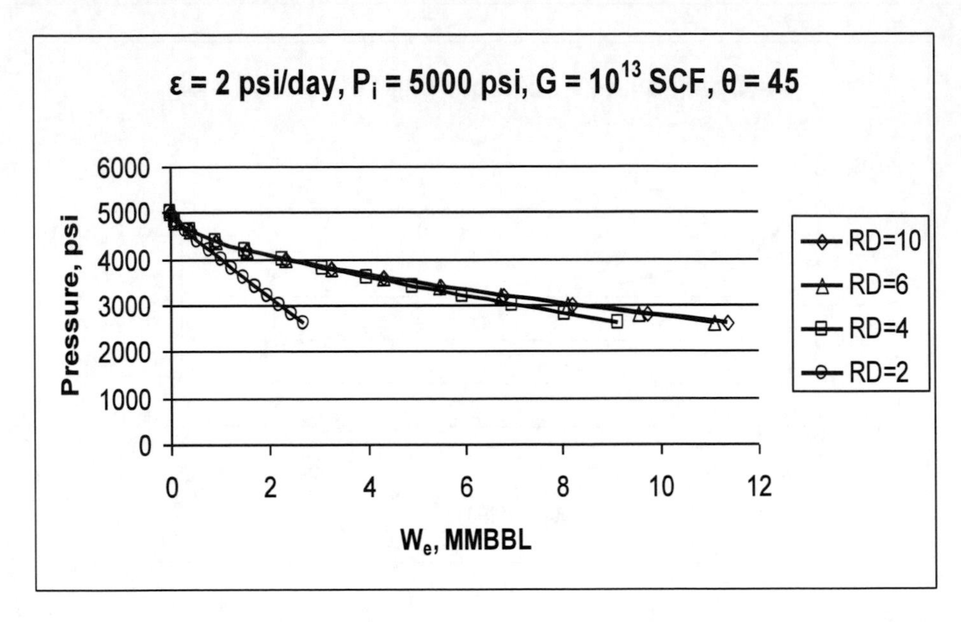

Figure 7. Effect of R_D on We; $G = 10^{13}$ scf.

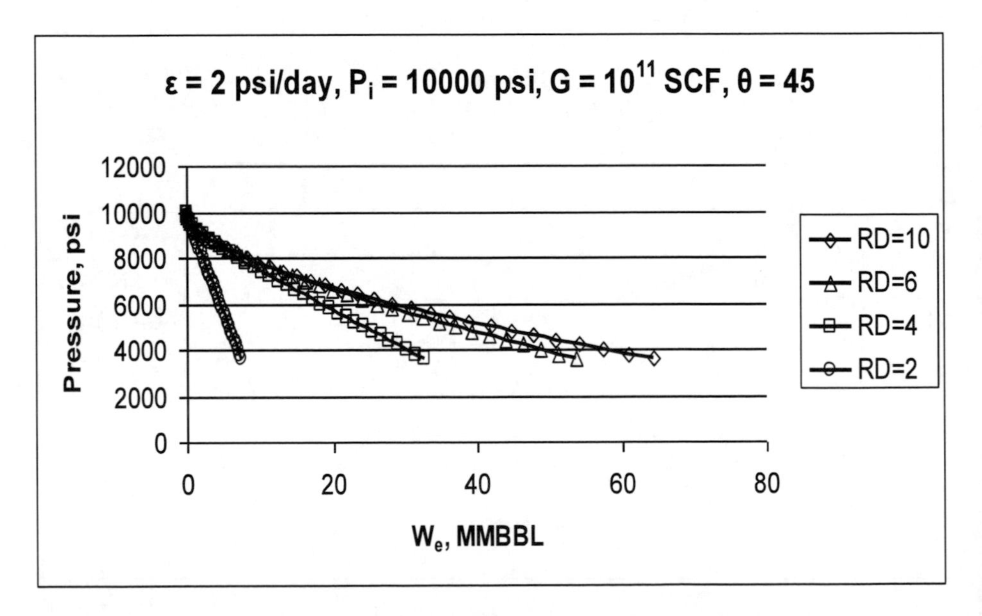

Figure 8. Effect of R_D on We; $p_i = 10{,}000$ psia.

A.2): Effect of aquifer size to reservoir size ratio (R_D) on *We* ($\theta = 180°$): The results are shown in Figure 6. The water encroachment angle enters the calculations in the aquifer constant (*U*). Consequently, as θ is increased from 45° to 180°, the results show significant increase in *We* (approximately by 4 folds).

A.3) Effect of aquifer size to reservoir size ratio (R_D) on *We* ($G = 10^{13}$ scf): The initial gas in place is increased to 10^{13} scf and the results of calculations are shown in Figure 7. At a given pressure and regardless of R_D, larger volume of gas in place yields approximately the same amount of water influx as the smaller one does. This may be attributed to the fact that R_D is simply a ratio between aquifer size to reservoir size, and increasing *G* would result in an increase in the aquifer size in the same proportion.

A.4) Effect of aquifer size to reservoir size ratio (R_D) on *We* (p_i = 10,000 psia): To investigate this effect, the initial reservoir boundary pressure is increased from 5,000 psia to 10,000 psia (doubled), and the results of calculations are presented in Figure 8. For R_D = 6, going from 10,000 psia down to 7,000 psia (reduction of 3,000 psia in p_i) results in 17 MMbbl of water influx compared with approximately the same amount of *We* in Figure 4 when p_i goes down from 5,000 psia to 2,000 psia. However, for R_D = 4, and for the same *pi* reduction, *We* values are 16 MMbbl and 12.5 MMbbl for p_i of 10,000 psia and 5,000 psia, respectively. Thus, the effect of p_i seems to become significant for $R_D \leq 4$. Effects of reservoir and aquifer parameters on cumulative gas production (*Gp*). A base case is selected as a reference for comparisons, where R_D =4, p_i = 5,000 psia, $G = 10^{11}$ scf, and θ =45°; Figure 9.

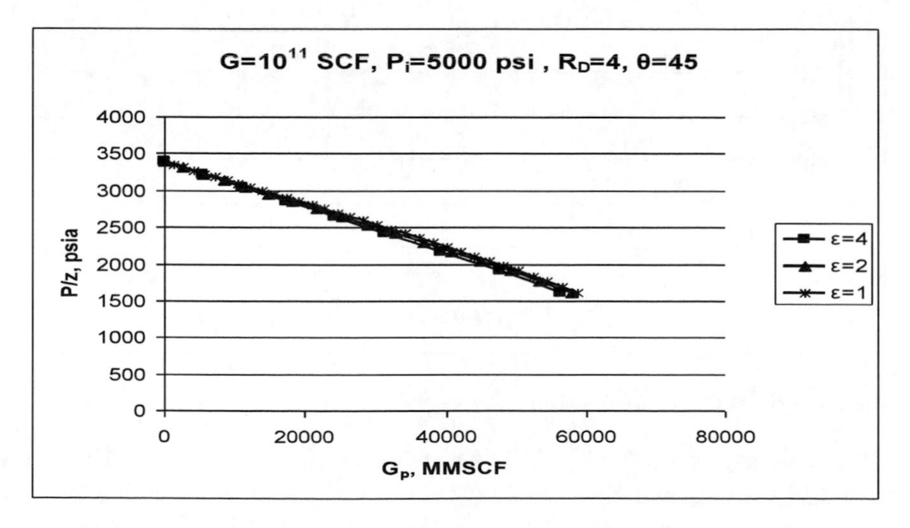

Figure 9. Effect of ε on *Gp*; base case, R_D = 4.

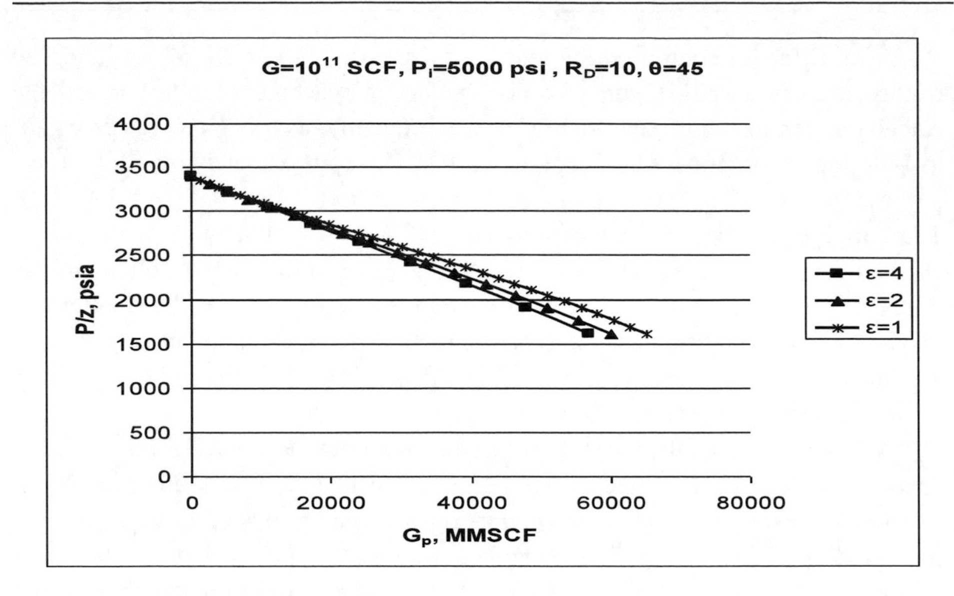

Figure 10. Effect of ε on Gp; $R_D = 10$.

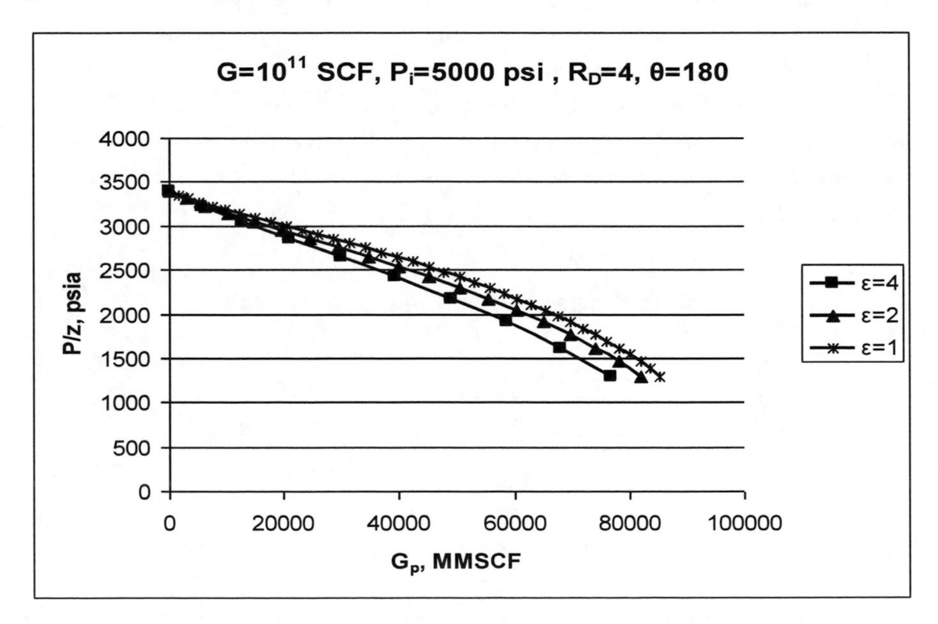

Figure 11. Effect of ε on Gp; $\theta. = 180^{o}$.

B.1) Effect of ε on Gp ($R_D = 10$): The results are shown in Figure 10 and they reveal that low values of ε tend to keep the p/Z vs. Gp curves at high level, while high ε permits drawing down the gas reservoir pressure before water influx completely floods the reservoir. Moreover, as the aquifer size (R_D) is increased from 4 to 10, the *p/Z* vs. *Gp* curves depart from the straight line relationships for a volumetric gas reservoir and tend to be at higher levels. In fact, when $R_D \leq 4$ the effect of the aquifer on gas reservoir performance can be neglected (Figure 9).

B.2) Effect of ε on *Gp* ($\theta = 180°$): The results of performance calculations are presented in Figure 11. They indicate that increasing the angle of water encroachment from 45° to 180° tends to keep the *p/Z* vs. *Gp* curves at higher levels (higher pressure for the same *Gp*). This effect is attributed to the larger amount of water influx into the reservoir, as the angle of water encroachment increases.

B.3) Effect of ε on *Gp* ($G = 10^{13}$): The results are shown in Figure 12. Increasing the initial gas in placc from 10^{11} scf to 10^{13} scf is found to further diminish the effect of the aquifer on gas reservoir performance and the reservoir would behave volumetrically, regardless of the rate of the reservoir boundary pressure depletion.

B.4) Effect of initial reservoir boundary pressure on *Gp*: The results of reservoir performance calculations are presented in Figures 13 and 14 for $R_D = 4$ and $R_D = 10$, respectively. For a given ε, when p_i is doubled the *p/Z* vs. Gp curves tend to be at higher levels, resulting in higher abandonment reservoir pressure.

***Part II**:* Sensitivity of gas recovery factor (*F*), abandonment reservoir pressure (p_{abn}), and number of lbm-moles of remaining gas in the reservoir (*n*) to the rate of reservoir boundary pressure depletion (ε)-case of infinite aquifers ($R_D \geq$ 10).

A base case is selected as a reference for comparisons where $G = 10^{11}$ scf, p_i = 5,000 psia, volumetric sweep efficiency (α) = 0.6, as shown in Figs. 15 and 16. The results of the base case reveal that the recovery factor tends to be lower at lower values of ε, which is indicative of the strong dependency of gas recovery on field gas offtake rate. Higher rates of boundary reservoir pressure depletion would result in lower abandonment pressures, higher recovery factors, and less number of lbm-moles of remaining gas in the reservoir.

The effect of increasing the volumetric sweep efficiency from 0.6 to 0.8 on the gas reservoir performance curves are shown in Figs. 17 and 18, respectively. For a given value of ε, a higher recovery factor, a lower p_{abn}, and a lower n are obtained compared with those in the base case. The reason for this improvement in the overall gas performance may be attributed to less amount of gas being left behind the water front at the higher volumetric sweep efficiency.

The effect of doubling the initial reservoir boundary pressure is shown in Figs. 19 and 20, respectively. For a given ε, a lower recovery factor, a higher p_{abn}

and a higher n are obtained compared with the base case. This may be attributed to the larger amount of water influx at the higher pressure than that at the lower pressure.

Part III: Sensitivity of F, $pabn$, and n to R_D-case of finite aquifers ($R_D < 10$). A base case is selected here for comparisons; $\alpha = 0.6$, $p_i = 5{,}000$ psia, $G = 10^{11}$ psia. The results of this base case are presented in Figs. 21 and 22. The recovery factor is very high and practically insensitive to R_D, but p_{abn} and n seem to be significantly affected when $R_D \leq 4$. This may be attributed to the fact that when $R_D \leq 4$, the gas reservoir behaves volumetrically, resulting in lower p_{abn}.

The effect of increasing α on the above parameters is shown in Figs. 23 and 24. Similar trends to those of the base case are observed, and as expected, they show better overall performance characteristics when compared with the base case.

Finally, the effect of increasing the initial gas in place to 10^{12} scf is shown in Figures 25 and 26. Similar trends and performance characteristics to the base case are observed.

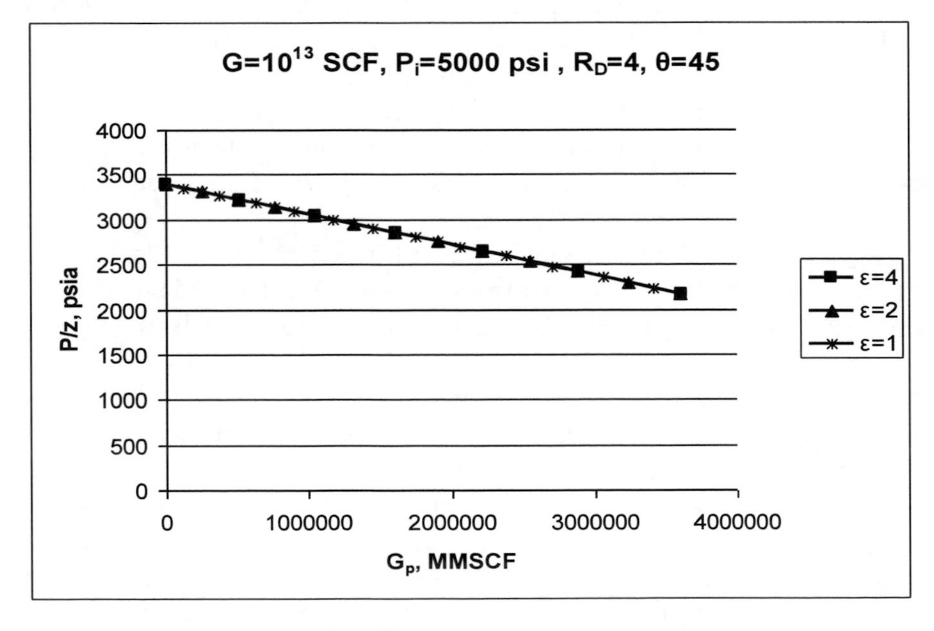

Figure 12. Effect of ε on Gp; $G = 10^{13}$ scf.

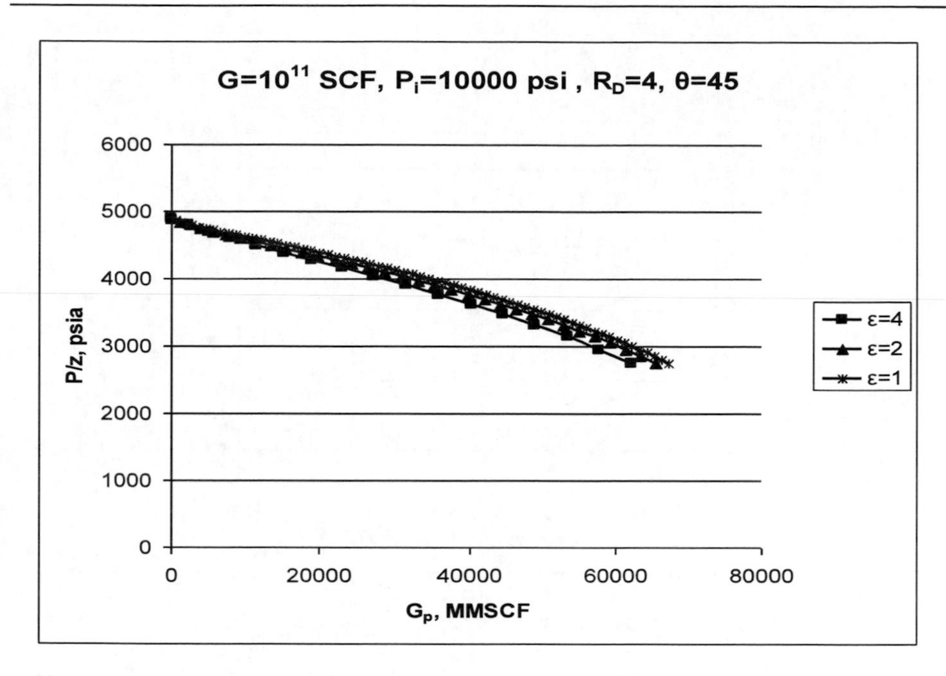

Figure 13. Effect of ε on Gp; $p_i = 10{,}000$ psia, $R_D = 4$.

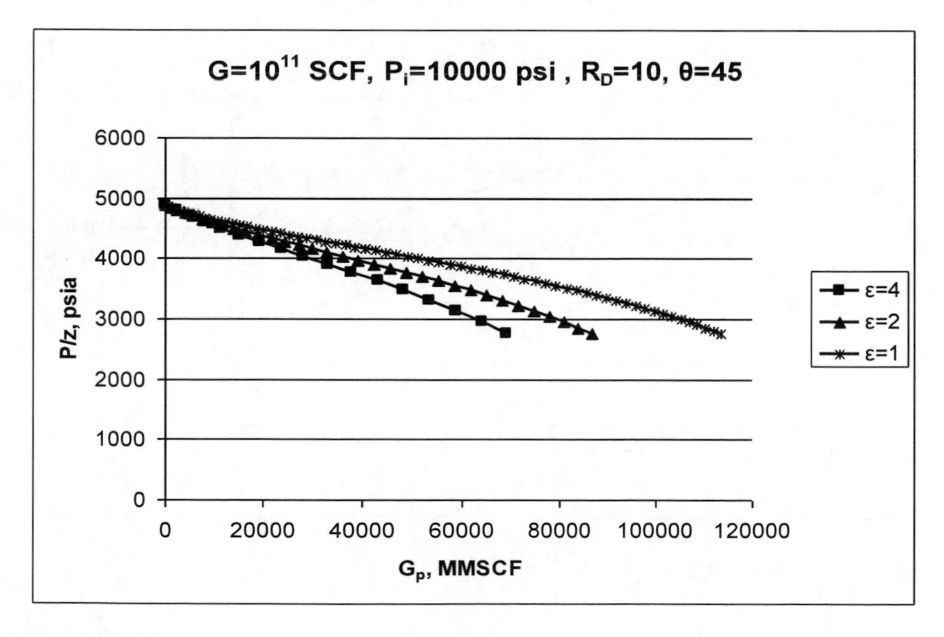

Figure 14. Effect of ε on Gp; $p_i = 10{,}000$ psia, $R_D = 10$.

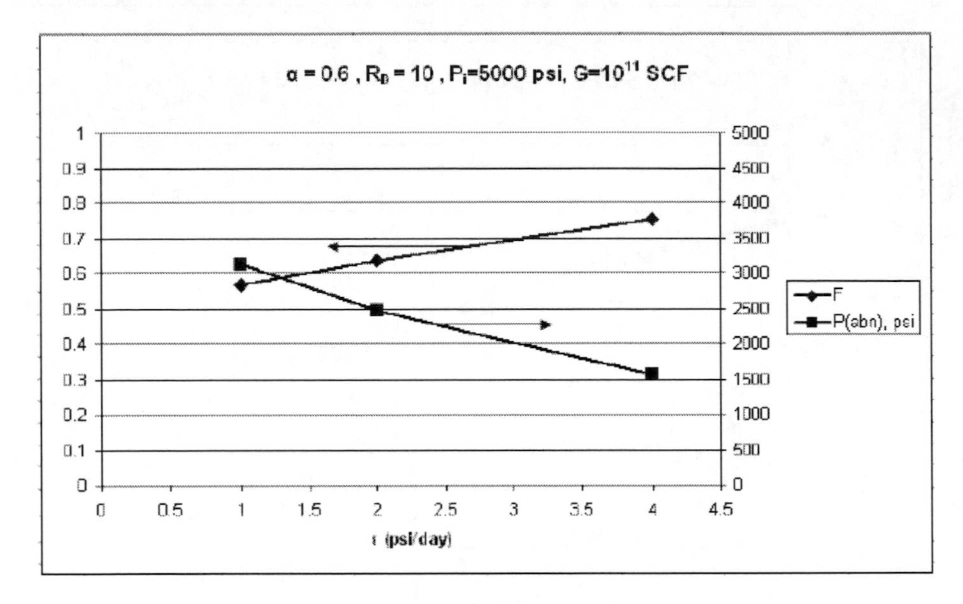

Figure 15. Sensitivity of F & p_{abn} to ε-infinite case.

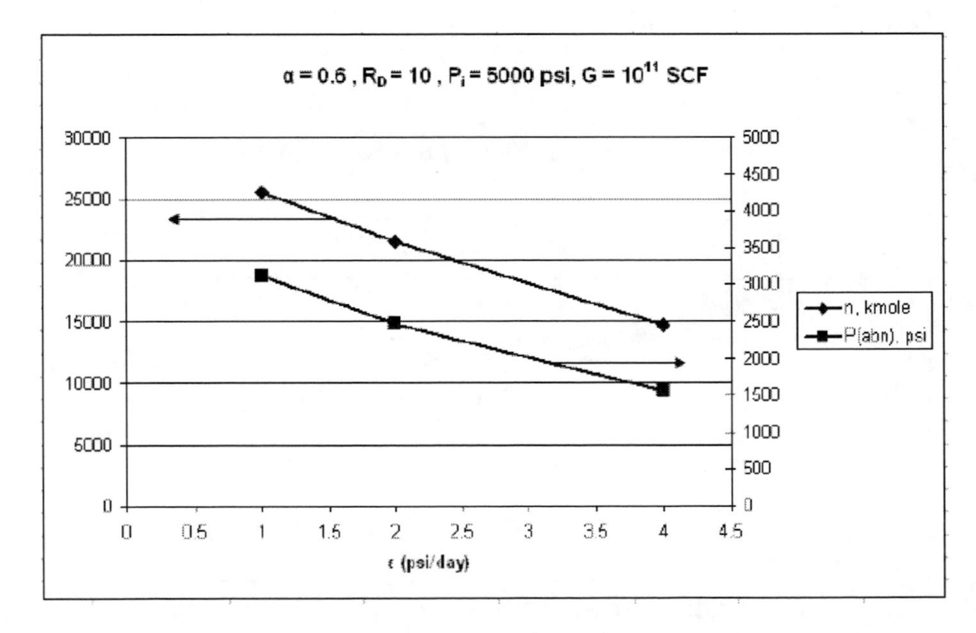

Figure 16. Sensitivity of p_{abn} & n to ε-infinite case.

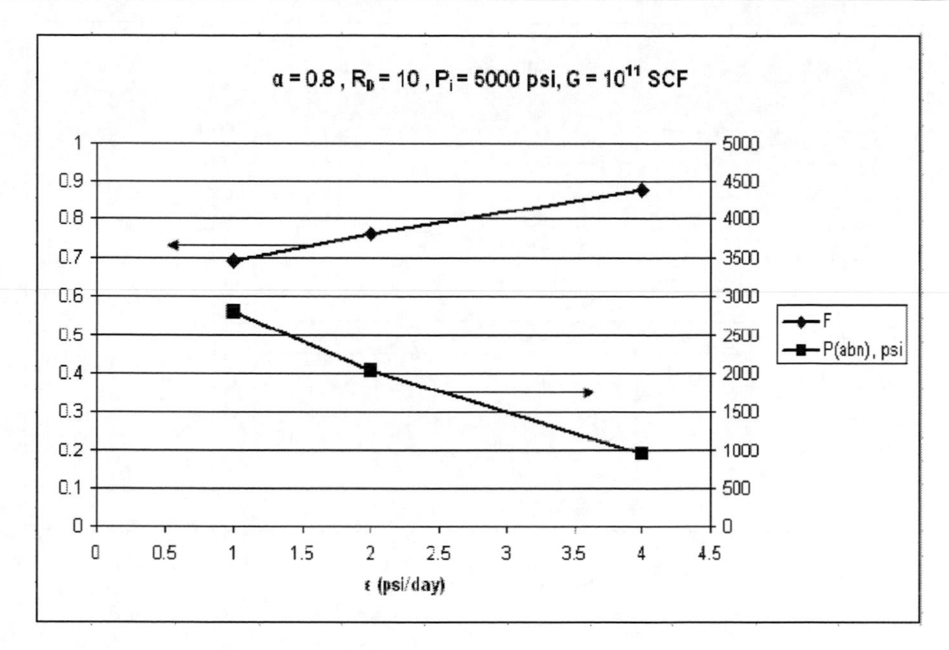

Figure 17. Sensitivity of F & p_{abn} to ε; $\alpha = 0.8$.

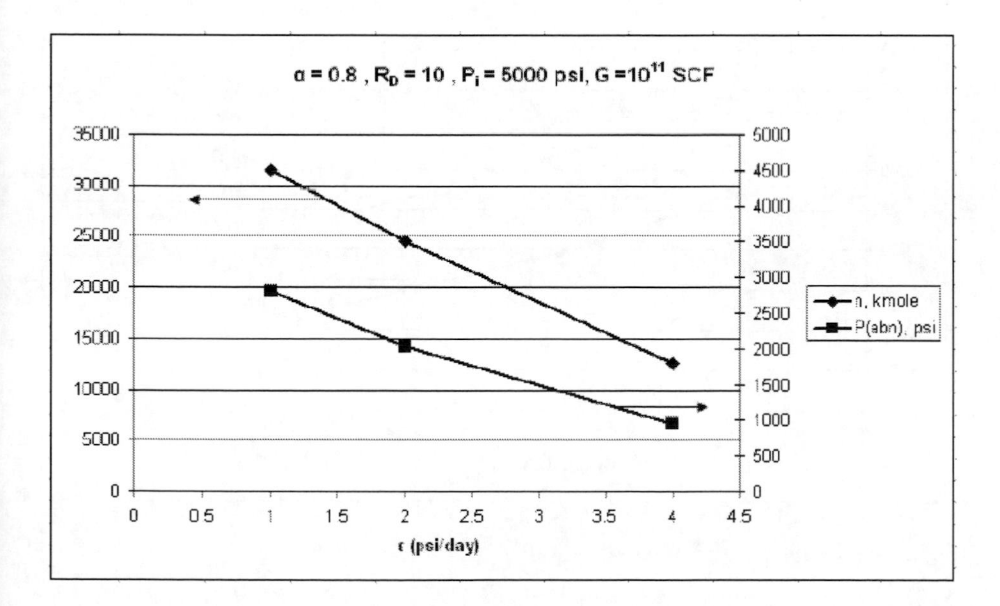

Figure 18. Sensitivity of p_{abn} & n to ε; $\alpha = 0.8$.

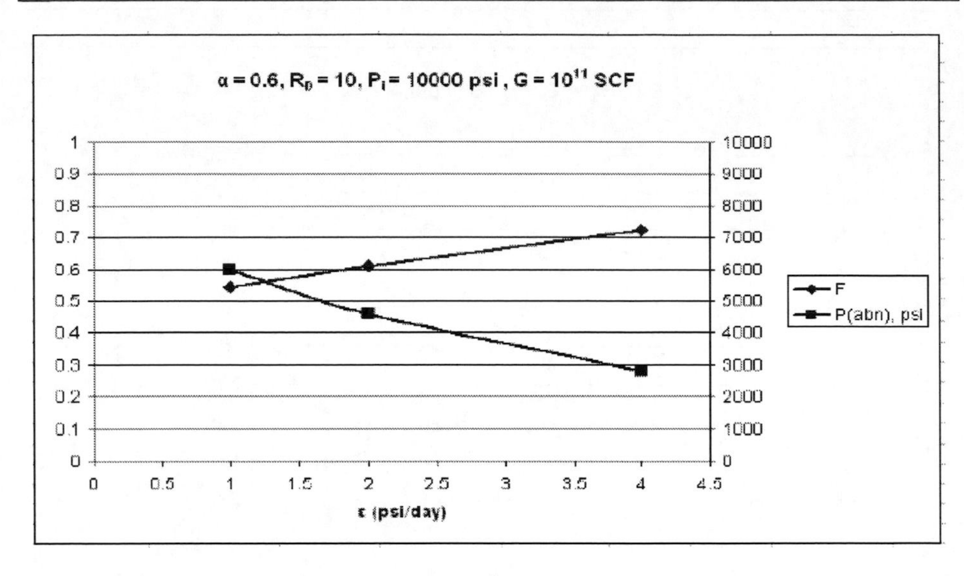

Figure 19. Sensitivity of F & p_{abn} to ε; $p_i = 10^4$ psi.

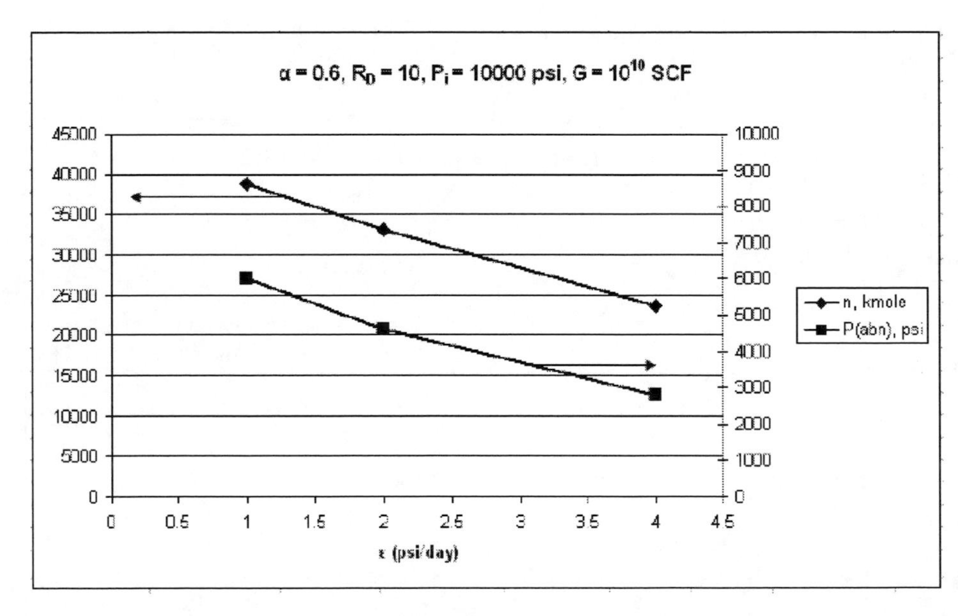

Figure 20. Sensitivity of F & n to ε; $p_i = 10^4$ psi.

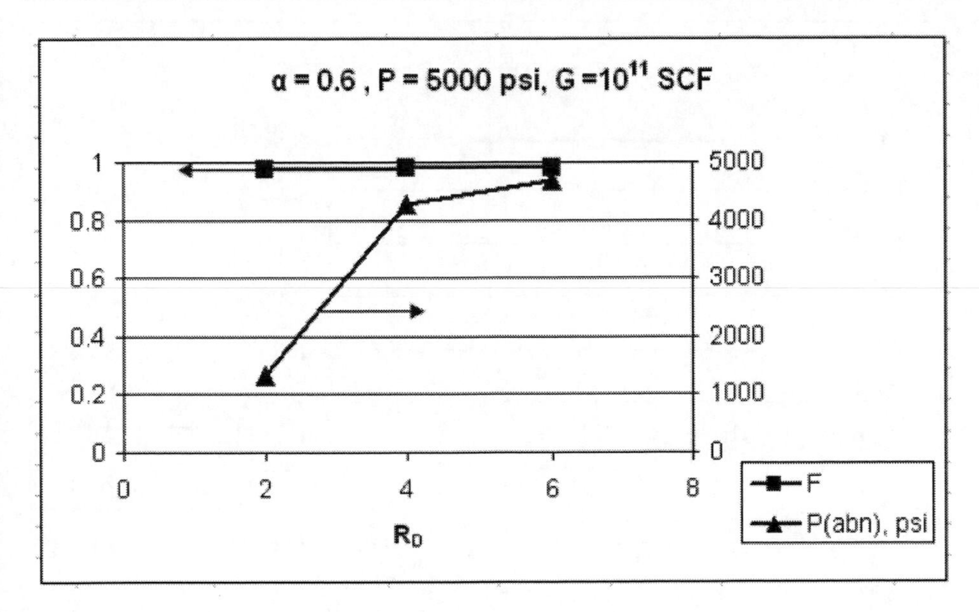

Figure 21. Sensitivity of F & p_{abn} to R_D-F.A

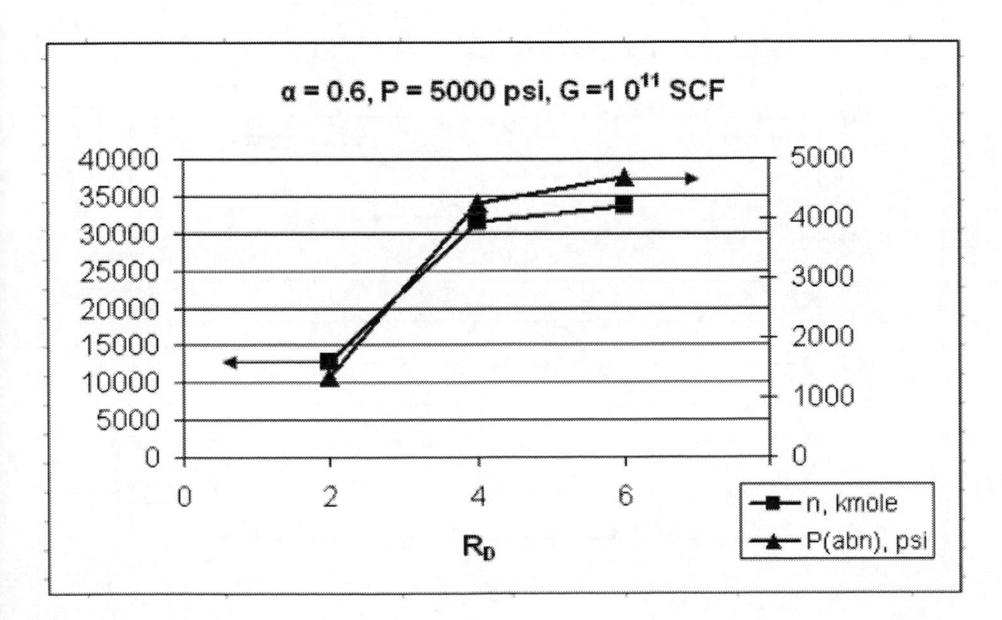

Figure 22. Sensitivity of n & p_{abn} to R_D-F.A.

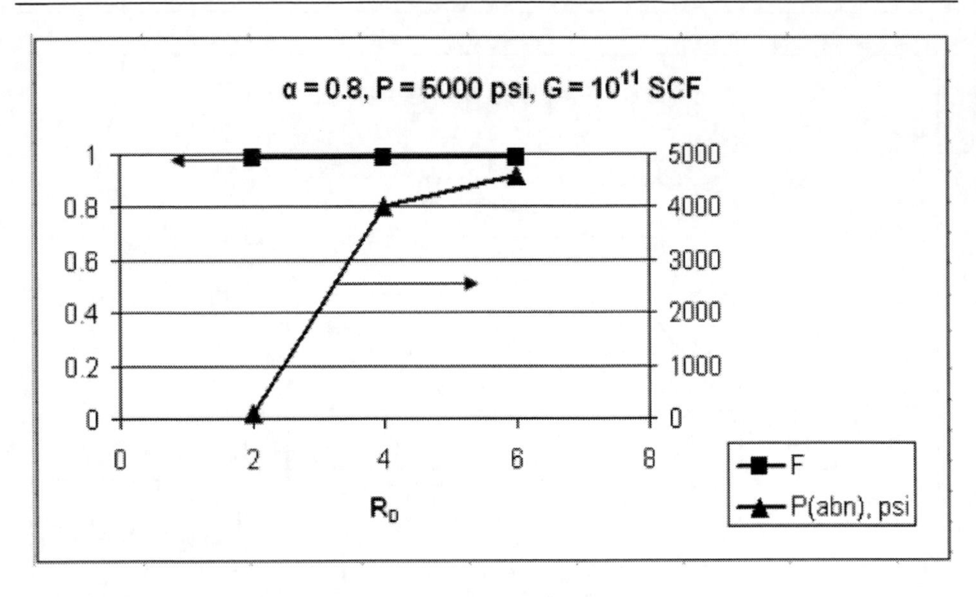

Figure 23. Sensitivity of F & p_{abn} to R_D-α = 0.8.

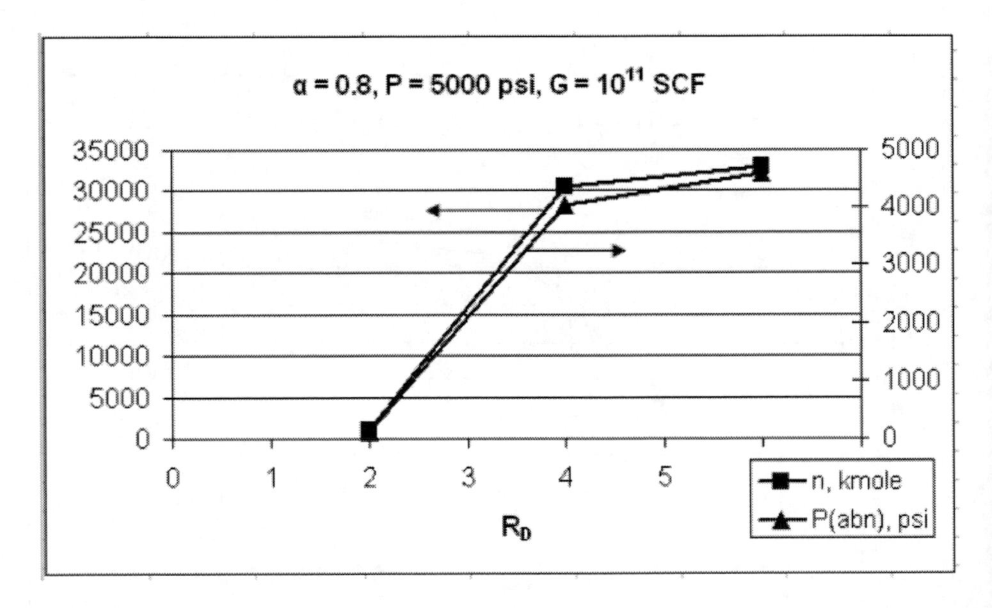

Figure 24. Sensitivity of n & p_{abn} to R_D-α = 0.8.

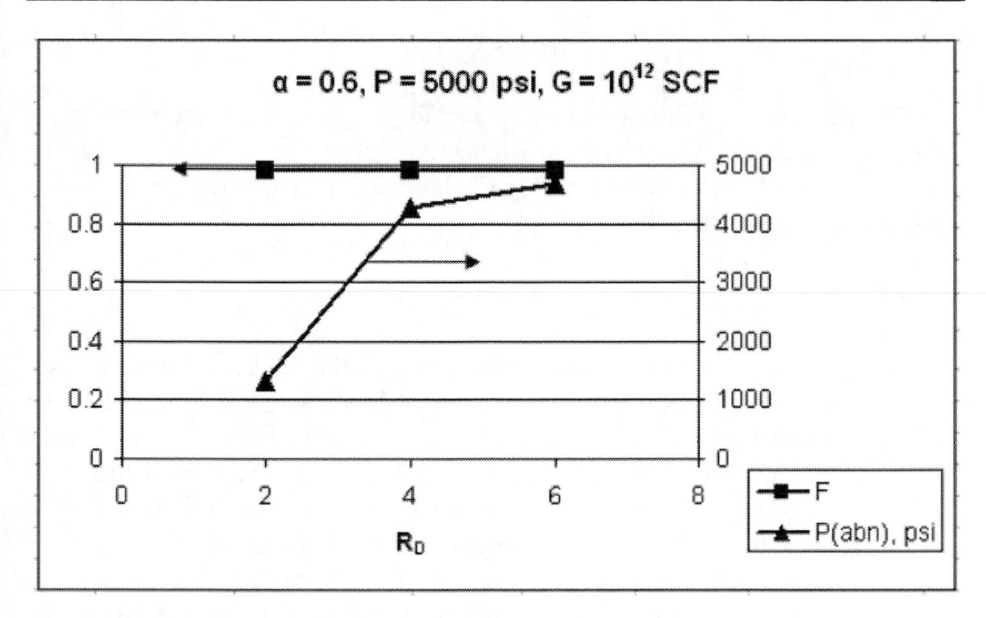

Figure 25. Sensitivity of F & p_{abn} to R_D;$G=10^{12}scf$.

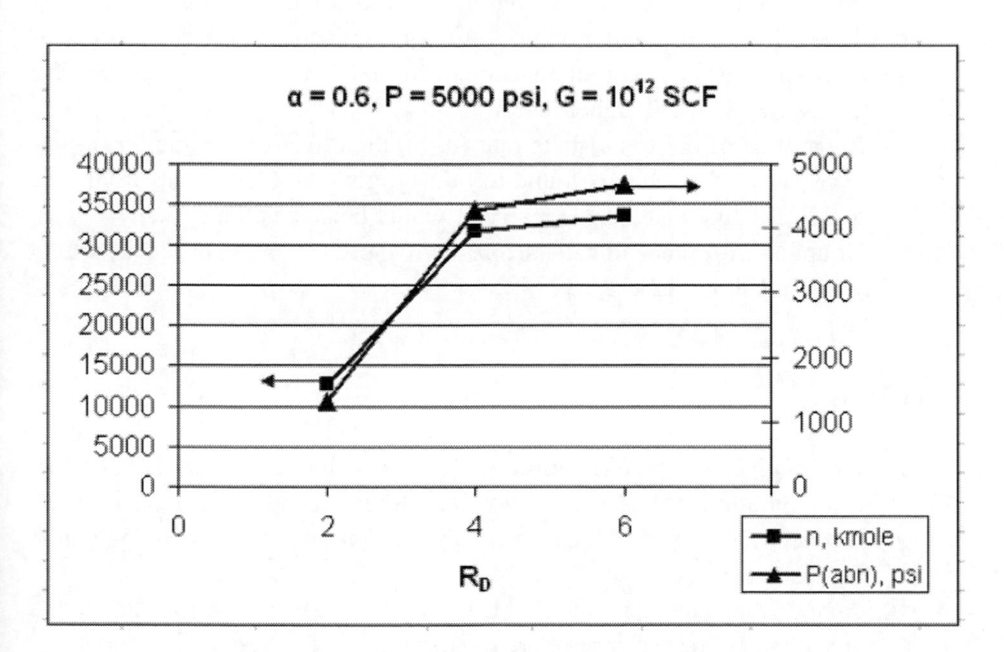

Figure 26. Sensitivity of n & p_{abn} to R_D;$G=10^{12}scf$.

CONCLUSION

On the basis of the results of this study for a hypothetical reservoir/aquifer case, the following conclusions are reached. These conclusions should provide the practicing engineer with some useful guide lines when developing gas reservoirs subjected to edge-water influx.

Part I

1. When $R_D \leq 2$ (small aquifer), the effect of the aquifer on gas reservoir performance becomes increasingly insignificant and may even be neglected.
2. Water influx is extremely sensitive to the angle of water encroachment.
3. At a given reservoir/aquifer boundary pressure and regardless of R_D, the water influx is found to be insensitive to the initial gas in place.
4. The effect of p_i on We seems to become significant when $R_D \leq 4$.
5. Low rates of gas offtake tend to keep the p/z vs. Gp curves at high level, while high gas offtake rates permit drawing down the gas reservoir pressure before water influx completely floods the reservoir. When $R_D \leq 4$ the effect of the aquifer on gas reservoir performance may be neglected.
6. Increasing the angle of encroachment from 45° to 180° tends to keep the p/z vs. Gp curves at higher levels.
7. Regardless of the gas offtake rate (or ε), the effect of the aquifer on gas reservoir performance is found to be negligible as G increases from 10^{11} to 10^{13} scf, and that the gas reservoir would behave volumetrically.
8. Doubling of p_i tends to shift the p/z vs. Gp curves to higher levels, which would result in higher P_{abn}

Part II

9. For a given volumetric sweep efficiency, higher rates of boundary reservoir/aquifer pressure depletion is found to decrease p_{abn}, increase recovery factor, and reduce the number of *lbm-moles* of remaining gas in the reservoir.
10. A new graphical approach has been proposed in this work to estimate the abandonment reservoir pressure for Infinite aquifers.

Part III

11. For a given volumetric sweep efficiency, the recovery factor seems to be very high and practically insensitive to R_D. However, P_{abn} and n appear to be significantly affected when $R_D \leq 4$.
12. A new graphical approach has been proposed to estimate the pressure depletion at the reservoir/aquifer boundary in finite aquifers.

REFERENCES

Agarwal, R.G., Al-Hussainy, R., and Ramey, H.J.Jr.:"The importance of Water Influx in Gas Reservoirs," *JPT* (Nov. 1965) 1336-1342, Trans., *AIME*, 234.

Al-Hashim, H.S., and Bass, D.M. Jr.:"Effect of Aquifer Size on the Performance of Partial Waterdrive Gas Reservoirs," *SPERE* (May 1988), 380-386.

Bruns, J.R., Fetkovich, M.J., and Meitzen, V.C.:"The Effect of Water Influx on p/z-Cumulative Gas Production Curves," *JPT* (March 1965) 287-295.

Dake, L.P. *Fundamentals of Reservoir Engineering,* Elsevier, Amsterdam, 1978.

Geffen, T.M. et al.:"Efficiency of Gas Displacement from Porous Media by Liquid Flooding," *Trans., AIME* (1952) 195, 29-35.

Havlena, D. and Odeh, A.S.:"The Material Balance as an Equation of a Straight Line," JPT (Aug. 1963) 896-900, *Trans., AIME.*

John Lee and Robert A. Wattenbarger:"Gas Reservoir Engineering," *SPE Textbook Series* Vol.5, SPE, Third Printing 2004, Appendix J.

Klins, M.A., Bouchard, A.J., and Cable, C.L.:"A Polynomial Approach to the van Everdingen-Hurst Dimensionless Variables for Water Encroachment," *SPERE* (Feb. 1988) 320-326.

Knapp, R.M. et al.:"Calculation of Gas Recovery Upon Ultimate Depletion of Aquifer Storage," *JPT* (Oct. 1968) 1129-1132.

Pepperdine, L.:"The Recognition and Evaluation of Water Drive Gas Reservoirs," Paper 78-29-38 presented at the *1978 Petroleum Soc. Of CIM Annual Technical Meeting.*

Shagroni, M.A.:"Effect of Formation Compressibility and Edge Water on Gas Field Performance," M.Sc. Thesis, Colorado School of Mines, Golden, CO. (1977).

van Everdingen, A.F. and Hurst, W.:"The Application of the Laplace Transportation to Flow Problems in Reservoirs," *Trans., AIME* (1949), 186, 305-324.

William, C. Lyons: *Standard Handbook of Petroleum and Natural Gas Engineering*, Vol.2, Gulf Publishing Co., Texas, 1996.

In: Natural Gas Systems
Editor: Rafiq Islam
ISBN: 978-1-61324-158-5

Chapter 2

FUZZY ESTIMATION AND STABILIZATION IN GAS LIFT WELLS BASED ON A NEW STABILITY MAP

E. Jahanshahi*[*]*, K. Salahshoor and Y. Sahraie
Department of Automation and Instrumentation,
Petroleum University of Technology, Tehran, Iran

ABSTRACT

System dynamics stability has been used to construct stability maps for gas-lifted oil wells. A fuzzy TS model is constructed based on proposed stability maps and Fuzzy Kalman Filter concept is used to estimate immeasurable state variables in noisy environment of well operation. Then, Fuzzy Linear Quadratic Gaussian regulator is constructed by combination of fuzzy estimator and fuzzy TS controller to stabilize the highly oscillatory well operations. The quadratic stability of the overall closed-loop system is evaluated in terms of Lyapunov with a less conservative criterion. Simulations are performed using a transient multi-phase flow simulator and variable input command signals are introduced to the system. Satisfactory results are obtained in terms of estimation and stabilization.

[*] Corresponding author. Email: e.jahanshahi@controlsys.net

Keywords: gas lift, casing heading instability, down-hole soft-sensing, Fuzzy Kalman Filter (FKF), Fuzzy Linear-Quadratic Gaussian (FLQG) regulator, Lyapunov stability.

NOMENCLATURE

g = Gravity constant, 9.81 m/s^2

R = Gas constant, 8314.51 J/kmol/°K

M = Gas molar weight, 16 gram/mol

ρ_o = Oil density, 781 kg/m^3

υ_o = Specific volume of oil, 0.00128 m^3/kg

P_r = Reservoir pressure, 1.5×10^7 pa

P_s = Separator pressure, 1×10^6 pa

T_a = Annulus temperature, 303 °K

T_t = Tubing temperature, 400 °K

V_a = Annulus volume, 37.68 m^3

V_t = Tubing volume, 7.584 m^3

L_t = Tubing length, 2400 m

L_r = Distant between reservoir and injection point, 150 m

A_t = Cross section area of tubing above injection point, 0.00316 m^2

A_r = Cross section area of tubing below injection point, 0.0031 m^2

C_{pc} = Production choke constant, 2×10^{-3}

C_{iv} = Injection orifice constant, 2×10^{-4}

ρ_m = Density of two phase fluid in well-head, kg/m^3

w_{iv} = Mass flow rate of injected gas into the tubing, kg/s

w_{pc} = Total mass flow rate through production choke, kg/s

w_{pg} = Mass flow rate of gas through production choke, kg/s

w_{po} = Mass flow rate of oil through production choke, kg/s

w_r = Mass flow rate of oil from reservoir, kg/s

$\rho_{a,i}$ = Gas density at injection point, kg/m^3

$P_{a,i}$ = Annulus pressure at injection point, pa

$P_{t,i}$ = Tubing pressure at injection point, pa

$P_{t,b}$ = Tubing pressure at bottom-hole, pa

P_t = Well-head pressure in tubing, pa

$u(t)$ = Production choke opening portion, $u(t) \in [0,1]$

INTRODUCTION

Because of drastic rises in crude oil price, use of new technologies for improved use of existing resources has gained considerable interest in these years. In this way down-hole soft sensing and stabilization of gas-lifted oil wells with highly oscillatory production flow rates has been topic of some recent researches (Jansen et al., 1999, Dalsmo et al., 2002). Figure 1 (Sinegre, 2006) shows a typical diagram of a gas-lifted oil well. Gas-lift is a method for activation of low pressure oil wells. In this method, gas is routed through surface gas injection choke (A) into the annulus (B) and then it is injected (C) into tubing (D) in order to be mixed with the fluid form reservoir (F). This reduces the density of oil column in tubing and lightens it, hence the production (E) rate from the low pressure reservoir is increased (Brown, 1982).

The oil production in the gas-lifted oil wells at their decline stages becomes unstable for low gas lift rates. There are several different phenomena to account for instability behaviour in wells (Hu and Golan, 2003). This study focuses on the instability of gas-lifted wells due to casing heading phenomenon. This instability may lead to periods of reduced or even no liquid production followed by large peaks of liquid and gas (Jansen et al., 1999). Figure 2 demonstrates a typical example of the casing heading phenomenon simulated in OLGA®v5.0 (Scandpower, 2006). The cyclic operation consists of three main phases (Sinegre et al., 2006) as follows:

- The upstream pressure is smaller than $P_{t,i}$, therefore no gas enters the tubing. The annulus pressure builds up until it reaches $P_{t,i}$. Then, injection into the tubing starts.
- As gas mixes with oil in the tubing, the column of liquid inside tubing lightens and the well starts producing. The gas injection rate does not fulfill the well's need. Therefore the pressure in the casing drops. Production reaches a maximum.
- Annulus pressure drops leading to reduction of the injection gas rate w_{iv} and the oil production. Less gas being injected, the oil column gets heavier and $P_{t,i}$ exceeds the upstream pressure. Gas injection in the tubing stops.

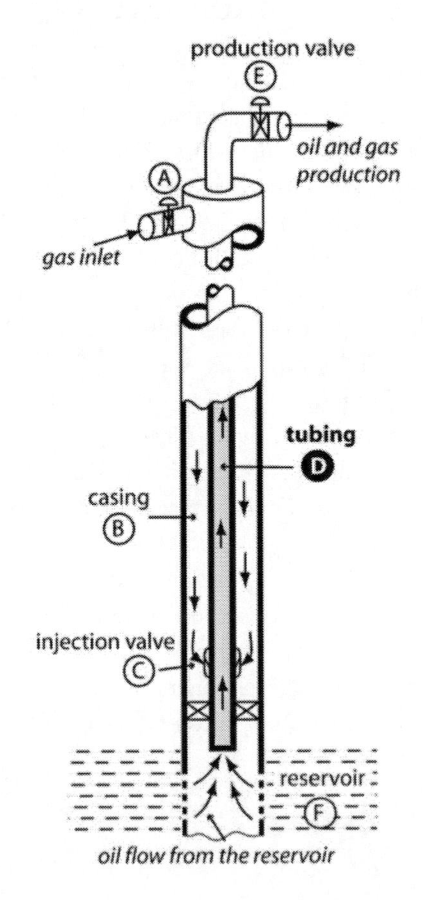

Figure 1. A gas lifted oil well.

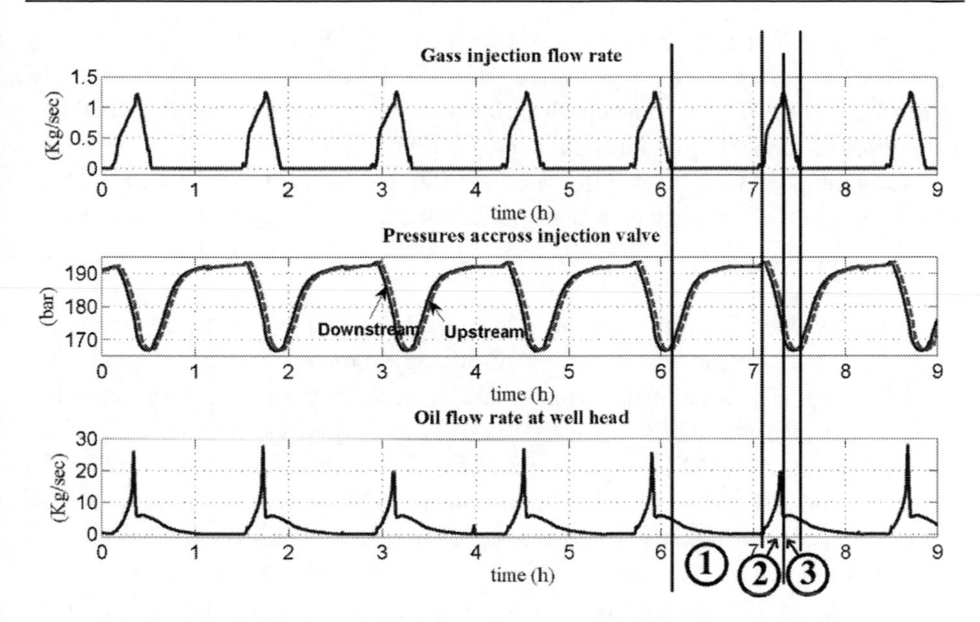

Figure 2. Casing-heading phenomenon simulated with OLGA®v5.0.

In order to suppress this oscillatory behaviour use of automatic feedback control has been considered and its advantages versus traditional methods have been addressed in (Jansen et al., 1999) and it has been investigated that substantial production improvement can be achieved by applying control to the gas lift system (Eikrem et al., 2002). State space model, based on mass balance equations, has been used and nonlinear full-state feedback has been applied for stabilization of the system (Imsland et al., 2003). But, some of these state variables are not measurable, therefore concept of state estimation from well-head measurements has been considered. Nonlinear observer is used for state estimation (Aamo et al., 2004) which has shown satisfactory result in experiment (Aamo et al., 2005). This concept has also been used for anti-slug control of gas lift wells experimentally (Eikrem et al., 2004a). The standard Kalman filter has been used for state estimation and down-hole soft-sensing based on locally linerarized model of the system (Eikrem et al., 2004b). Advantage of estimators based on Kalman Filter compared to nonlinear observer is capability of working in presence of process and measurement noises (Simon, 2006). But, the standard Kalman filter could be used only in a single operating point for a locally linearized dynamic of the system. To deal with this problem, extended Kalman Filter (EKF) has been proposed in (Bloemen et al., 2004) as a soft-sensing technique for estimation of down-hole variables that are not directly measurable.

This work is an extension of the work done in (Eikrem et al., 2004b) using universal approximation capabilities of TS fuzzy systems (Passino and Yurkovich, 1998). This concept has been used in design of fuzzy Kalman estimators (Simon, 2003, Layne and Passino, 1996).

Combination of Kalman Filter for state estimation and linear-quadratic (LQ) state-feedback for state-space system is referred to as linear-quadratic Gaussian (LQG) structure. This structure is also exists for TS fuzzy models (Ma et al., 1998, Simon, 2003).

Stability maps (Fairuzov et al., 2004) are proposed for using in the design step of gas lift systems and later for defining appropriate operating strategies for gas-lift systems. These maps usually are generated based on some stability criteria. Many such stability criteria has been proposed (Poblano et al., 2005, Alhanati et al., 1993). This paper proposes a stability map based on system dynamics stability. Also, we used the proposed stability map for construction of fuzzy sets that are used in fuzzy estimator and fuzzy LQG regulator.

Open-loop model is constructed in OLGA®v5.0 and fuzzy LQG regulator is simulated in MATLAB, and two parts are connected using Matlab-Olga Link® Toolbox.

The organization of this paper is as follows. At first, the mathematical model of system is described. The next section introduces construction of the stability maps. Then, a fuzzy Kalman estimator will be presented. Afterwards, an optimal fuzzy control strategy will be developed to stabilize the well oscillatory behaviour. Then, quadratic stability analysis of overall closed-loop system is evaluated with lyapunov stability criterion. Subsequently, the proposed controller will be tested on a case study. Finally, the results are summarized.

MATHEMATICAL MODEL

Many mathematical models have been proposed for gas-lifted oil wells. A very detailed and precise model based on laboratory data has been presented in (Terre et al., 1987) for casing-heading phenomenon. But, generally simplified versions of two-phase flow model have application in stability analysis and stabilization of gas lifted oil wells. The gas-lifted oil well operation can be described by following state-space equations (Aamo et al.):

$$\dot{x}_1 = w_{gc} - w_{iv} \tag{1}$$

$$\dot{x}_2 = w_{iv} - w_{pg} \tag{2}$$

$$\dot{x}_3 = w_r - w_{po} \tag{3}$$

where the state variables consist of x_1 as the mass of gas in the annulus, x_2 as the mass of gas in tubing, and x_3 as the mass of oil in tubing. The mass flow rates in the state-space model are given by:

$$w_{gc} = \text{Constant flow rate of lift gas.} \tag{4}$$

$$w_{iv} = C_{iv}\sqrt{\rho_{a,i}\max\{0, p_{a,i} - p_{t,i}\}} \tag{5}$$

$$w_{pc} = C_{pc}\sqrt{\rho_m \max\{0, p_t - p_s\}u} \tag{6}$$

$$w_{pg} = \frac{x_2}{x_2 + x_3} w_{pc} \tag{7}$$

$$w_{po} = \frac{x_3}{x_2 + x_3} w_{pc} \tag{8}$$

$$w_r = C_r(P_r - P_{t,b}) \tag{9}$$

Symbols and their nominal values for a case study (Sinegre et al.) are described in the nomenclature. Algebraic thermodynamic “equations of state” (EOS) (Sonntag et al., 2003) relating density, temperature and pressure in tubing and casing are as follows:

$$P_{a,i} = \left(\frac{RT_a}{V_a M} + \frac{gL_a}{V_a}\right)x_1 \tag{10}$$

$$P_t = \frac{RT_t}{M}\frac{x_2}{L_t A_t + L_r A_r - \upsilon_o x_3} \tag{11}$$

Also,

$$P_{t,i} = P_t + \frac{g}{A_t}(x_2 + x_3 - \rho_o L_r A_r) \tag{12}$$

$$P_{t,b} = P_{t,i} + \rho_o g L_r \tag{13}$$

Two densities used in above equations are defined as:

$$\rho_{a,i} = \frac{M}{RT_a} P_{a,i} \tag{14}$$

$$\rho_m = \frac{x_2 + x_3 - \rho_o L_r A_r}{L_t A_t} \tag{15}$$

Finally, state space model can be written as following:

$$\dot{x}_1 = w_{gc} - w_{iv} \tag{16}$$

$$\dot{x}_2 = w_{iv} - \frac{x_2}{x_2 + x_3} w_{pc} \tag{17}$$

$$\dot{x}_3 = w_r - \frac{x_3}{x_2 + x_3} w_{pc} \tag{18}$$

The mathematical model must be able to describe the interaction between annulus and tubing for different amounts of injected gas. Figure 3 shows behaviour of flow rate through production choke for three different levels of injected gas based on the derived model described by equations (16)-(18). As illustrated, the well production becomes unstable at low values of injected gas while as injected gas amount is increased the model shows stable behaviour that is consistent with some state of art knowledge about gas lifted oil wells.

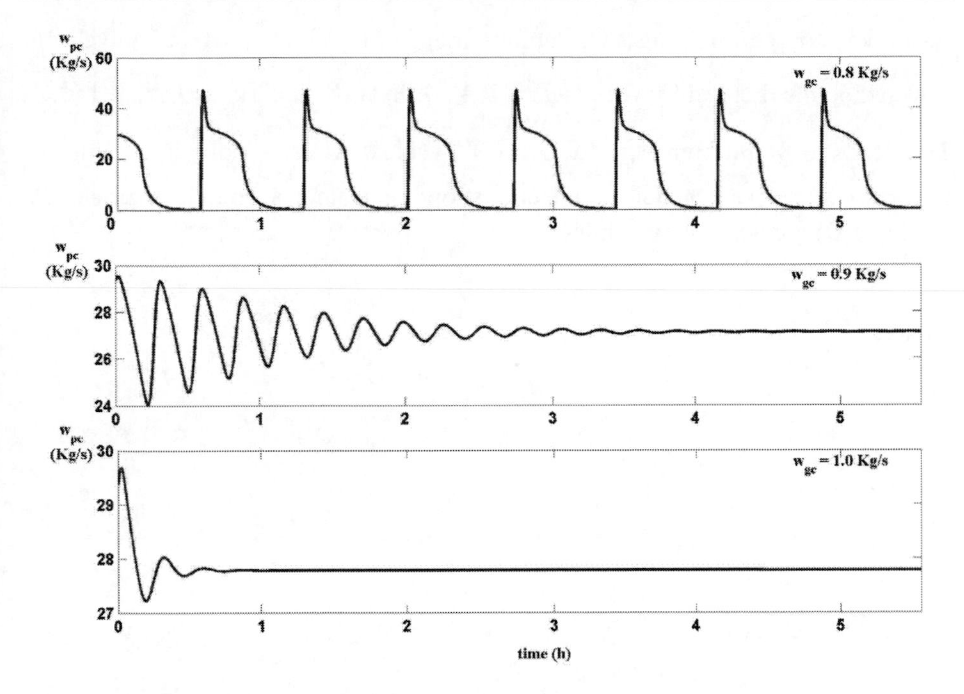

Figure 3. Production choke flow rate for 3 different values of injected gas.

A New Stability Map

Stability maps are usually used in designing step of system and later for defining appropriate operating strategies for gas-lift systems. Several techniques have been proposed in the literature to analyze gas-lift stability (Poblano et al., 2005).

In this work, a simple approach has been proposed to derive a stability map based on linearization of the nonlinear model defined by

$$\begin{cases} \dot{x} = f(x,u,w_{gc}) \\ y = g(x,u,w_{gc}) \end{cases} \tag{19}$$

in a set of equilibrium operating points. These operating points should be such that the overall well operating region be covered. To obtain a reasonable resolution for constructing the stability map, $f(x,\bar{u},\bar{w}_{gc})=0$ is solved for selected points dictated by vectors $\bar{u}=0.4:0.05:0.8$ and $\bar{w}_{gc}=0.4:0.1:1.2$. This leads to 81 different $\bar{x}=[\bar{x}_1,\bar{x}_2,\bar{x}_3]^T$ values. At each individual point, the linearized state space model and the corresponding state-space model matrices (A, B, C and D) are derived as follows:

$$A=\left.\frac{\partial f(x,u,w_{gc})}{\partial x}\right|_{x=\bar{x},u=\bar{u},w_{gc}=\bar{w}_{gc}} \tag{20}$$

$$B=\left.\frac{\partial f(x,u,w_{gc})}{\partial u}\right|_{x=\bar{x},u=\bar{u},w_{gc}=\bar{w}_{gc}} \tag{21}$$

$$C=\left.\frac{\partial g(x,u,w_{gc})}{\partial x}\right|_{x=\bar{x},u=\bar{u},w_{gc}=\bar{w}_{gc}} \tag{22}$$

$$D=\left.\frac{\partial g(x,u,w_{gc})}{\partial u}\right|_{x=\bar{x},u=\bar{u},w_{gc}=\bar{w}_{gc}} \tag{23}$$

Sate-space matrices also will be used in next section in construction of fuzzy model. Using this procedure, the eigenvalues and steady-state values for each point in the stability map can be calculated. This provides the necessary means to assess the stability condition of each operating point based on the sign of real-part of the calculated eigenvalues. The resulting stability map has been illustrated in Figure 4 which shows w_{pc} versus w_{gc} for different values of u. As indicated, stable points are marked with + and unstable points are marked with × for clarity. As expected, decreasing w_{gc} for a specified value of u, leads to destabilization of system.

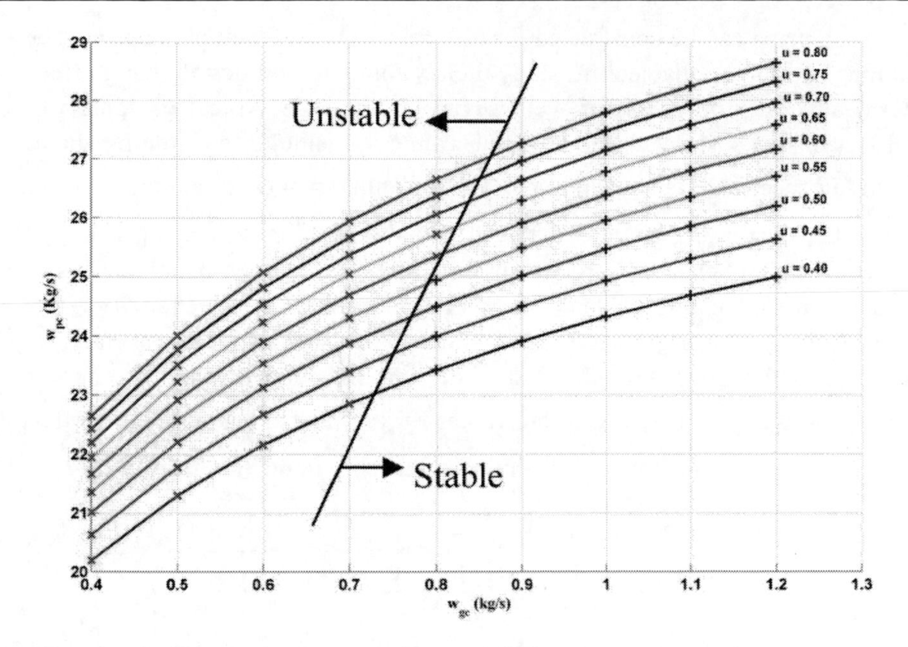

Figure 4. Production flow rate versus injected gas for different values of u.

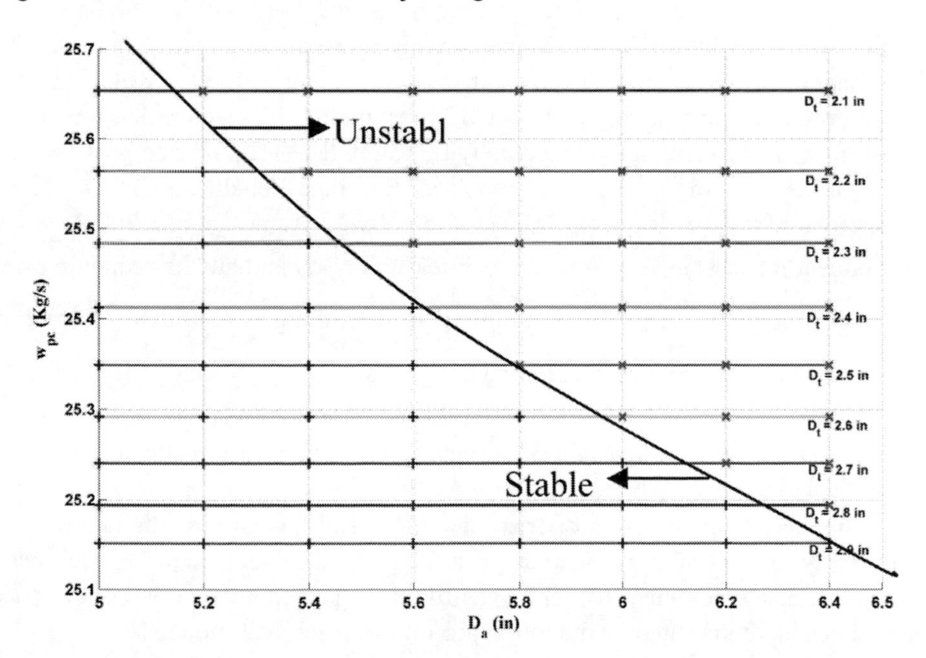

Figure 5. Production flow rate for specified values of wgc=0.8 kg/s and u=0.6 with different values of annulus and tubing diameter.

Such maps can be constructed for different values of well parameters such as annulus and tubing diameter to study their stabilizing and destabilizing effect on the steady state operation of well. Figure 5 shows a typical example of the stability map constructed for different values of annulus and tubing diameter assuming specified value of $w_{gc} = 0.8 kg / s$ and $u = 0.6$. Examining the figure shows that increasing annulus diameter destabilizes the system for steady state operation without changing production flow rate. This is a reasonable observation, because larger annulus volume means that a decrease in injection flow rate at surface leading to larger decrease in annulus pressure and consequently larger decrease in flow rate through bottom-hole gas injection orifice. In other hand, increasing tubing diameter for same values of w_{gc} and u has stabilizing effect in steady state operation, but causes a little decrease in production flow rate.

DESIGNING FUZZY ESTIMATOR

In practice, down-hole measurements relating to tubing and annulus variables are not in general available. If they can be measured, they will be unreliable due to the harsh conditions in which the measuring sensors operate. Thus, in this research study it is assumed that only the measurements at the top of the annulus and tubing are available and rest should be estimated. It is noted that the single phase flow in the annulus can accurately be estimated based on one pressure and one temperature measurement. Therefore, the main challenge concerns the estimation of the multiphase flow in the tubing (Aamo et al., 2004). As a consequent it is assumed that x_1 can be measured. To estimate the remaining two states (x_2 and x_3), the pressure at the top of the tubing is measured. Therefore, the available measurements are assumed to be $y_1 = x_1(t)$ and $y_2(t) = p_t$. In addition, it is assumed that the variables relating to opening portion of the production choke and flow rate of surface gas injection are available (Aamo et al.).

Nonlinear systems can be approximated as locally linear systems in much the same way that nonlinear functions can be approximated as piecewise linear functions. Thus, nonlinear model of gas-lifted oil presented as a 3-D state-space model can be represented by the following fuzzy local linear models:

$$\text{if } z[k] \text{ is } F_i \text{ then } x[k+1] = A_i x[k] + B_i u_d[k] + G_i w_d[k]$$
$$y[k] = C_i x[k] + H_i w_d[k] + v[k], \qquad i = 1,..,L \tag{24}$$

The $z[k]$ is vector of premise variables, $L = 81$ is the number of fuzzy rules, k is the time index, F_i are fuzzy sets, $x[k] \in R^n$ is the state vector, $u_d[k] \in R^m$ is the deterministic input, $w_d[k]$ is the process noise, $y[k] \in R^r$ is the measured output, and $v[k]$ is the measurement noise. We assume that the process noise $w_d[k]$ and measurement noise are white with known covariance values of $\mathrm{E}(w_d w_d^T) = Q$, $\mathrm{E}(vv^T) = R$ and $\mathrm{E}(w_d v^T) = N$. And $\mathrm{E}(w_d) = \mathrm{E}(v) = 0$, where E denotes Expected Value operator. The fuzzy combination of these local models results in the global model:

$$x[k+1] = \sum_{i=1}^{L} h_i(z[k])\{A_i x[k] + B_i u_d[k] + G_i w_d[k]\},$$
$$y[k] = \sum_{i=1}^{L} h_i(z[k])\{C_i x[k] + H_i w_d[k] + v[k]\} \tag{25}$$

where the membership grades $h_i(z[k])$ are defined as:

$$h_i(z[k]) = \frac{\mu_i(z[k])}{\mu[k]} \tag{26}$$

$$\mu[k] = \sum_{i=1}^{L} \mu_i(z[k]) \tag{27}$$

$$z[k] = [u_t[k] \; w_{gc,t}[k]]^T \tag{28}$$

The premise variables used in this fuzzy modelling are steady-state plus deviation values of u and w_{gc} ($u_t = \bar{u} + u_d$ and $w_{gc,t} = \bar{w} + w_d$), which are available at well-head. To fuzzify these premise variables, 2-D piecewise linear membership functions are used as shown in Figure 6. It's noted that centres of antecedent membership functions are same as points used in stability maps proposed in this paper.

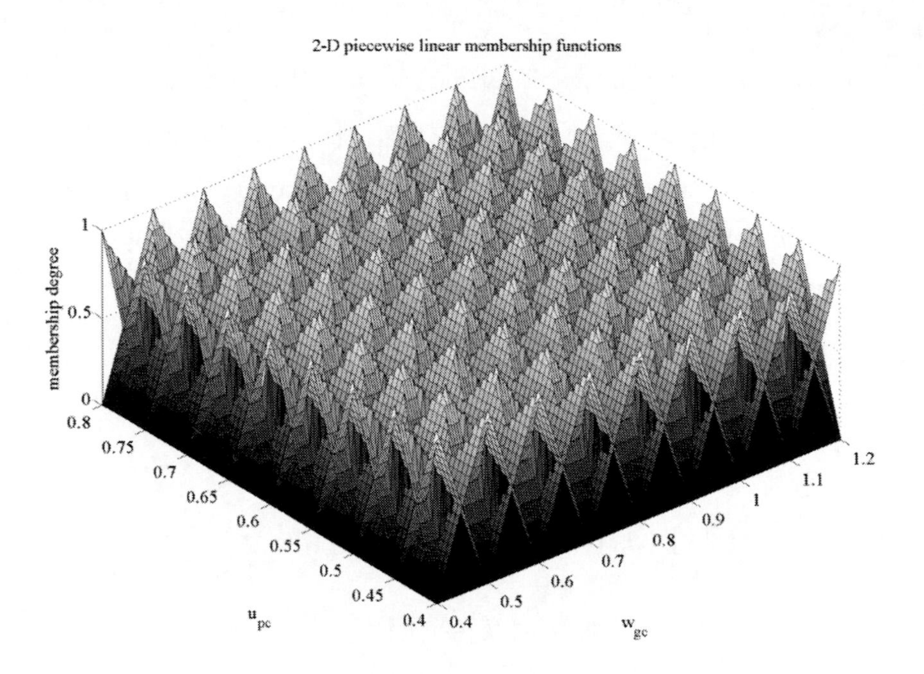

Figure 6. Membership functions in antecedent part of fuzzy system.

Definition 1: If the pairs $(A_i, C_i), i = 1,..., L$ are observable, the fuzzy system in equation (24) is called locally observable. For the fuzzy observer design, it is assumed that the fuzzy system is locally observable (Ma et al., 1998).

A standard Kalman filter (Simon, 2006) can be used as a state estimation algorithm for each local linear model. Consequently, the nonlinear system state estimation can be implemented using the fuzzy aggregation of local estimates as follows:

$$\hat{x}[k+1] = \sum_{i=1}^{L} h_i(z[k])\{A_e^i \hat{x}[k] + B_e^i u_e[k]\}; \tag{29}$$

$$\hat{y}[k+1] = \sum_{i=1}^{L} h_i(z[k])\{C_e^i \hat{x}[k]\} \tag{30}$$

where $A_e^i = A_i - L_i C_i$, $B_e^i = [B_i, L_i]$ and $C_e^i = C_i$ are state space matrices of Kalman filter for i'th rule and L_i is derived by solving a discrete Riccati equation (Burl, 1999). $u_e = [u_d \ y_v]^T$, where y_v denotes the noisy output measurement. u_d is treated as known input, while w_d is considered as unknown input (process noise). It's a reasonable assumption because deviations about steady state values are used for each linear model. Also, w_{gc} depends on line pressure and its fluctuations about steady-state are considered as Gaussian random noise.

FUZZY LQG REGULATOR DESIGN

As shown in Figure 7, the optimum region for production of gas lift system is upward region of production curve (Aamo et al., 2004, Jansen et al., 1999), because any increase in gas injection leads to production increase. These regions however are unstable as has been demonstrated in the proposed stability maps. Thus, the desired control scheme should be able to change the dynamic of system in unstable regions so as to achieve a stable operation in these regions. For this purpose, Linear-quadratic state-feedback regulator for each local state-space system is utilized in which K_i, the state-feedback gain in each rule, is derived by solving a discrete Riccati equation (Burl, 1999).

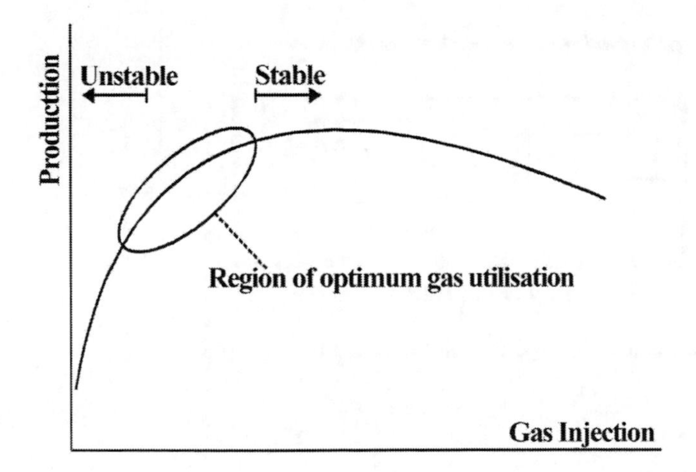

Figure 7. Production curve of a gas lift system.

Definition 2: If the pairs $(A_i, B_i), i = 1, \ldots, L$ are controllable, the fuzzy system in equation (24) is called locally controllable. For the fuzzy controller design, it is assumed that the fuzzy system is locally controllable (Ma et al., 1998).

Conjunction of the state-feedback with the Kalman estimator in each rule leads to a fuzzy LQG controller as below:

$$\hat{x}[k+1] = \sum_{i=1}^{L} h_i(z[k])\{A_R^i \hat{x}[k] + B_R^i y_v[k]\} \tag{31}$$

$$u_c[k+1] = \sum_{i=1}^{L} h_i(z[k])\{C_R^i \hat{x}[k]\} \tag{32}$$

where A_R^i, B_R^i and C_R^i are state space matrices of LQG regulator for i'th rule. Figure 8 shows control structure that is used in each rule of fuzzy system.

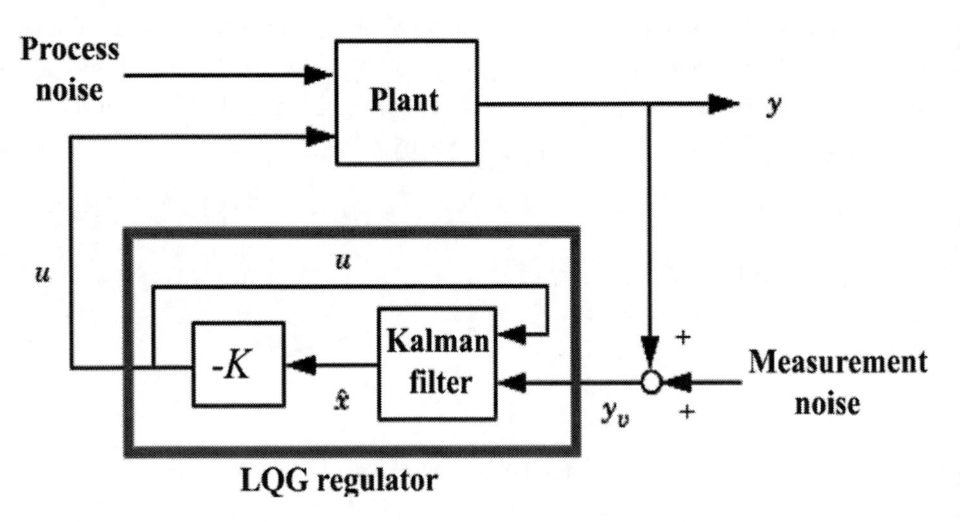

Figure 8. Structure of closed-loop system used for each rule.

CLOSED-LOOP STABILITY

For stability analysis, we use the direct method of Lyapunov. Choosing a quadratic Lyapunov function (Passino and Yurkovich, 1998):

$$V(x) = x^T Px \tag{33}$$

where P is a "positive definite matrix" (denoted by $P > 0$) that is symmetric (i.e., $P = P^T$). Given a symmetric matrix P, we can easily test if it is positive definite. You simply find the eigenvalues of P, and if they are all strictly positive, then P is positive definite. If P is positive definite, then for all $x \neq 0, x^T Px > 0$. Hence, we have $V(x) > 0$ and $V(x) = 0$ only if $x = 0$. Also, if $|x| \to \infty$, then $V(x) \to \infty$.

To show that the equilibrium $x = 0$ of the closed-loop system resulting from feedback connection of open-loop system and proposed regulator in equations (31)-(32) is globally asymptotically stable, we need to show that $\dot{V}(x) < 0$ for all x (Passino and Yurkovich, 1998). This condition holds, if there exists a common positive definite matrix P such that

$$(A_c^{ii})^T P + PA_c^{ii} < 0 \tag{34}$$

for $i = 1, \ldots, L$, and

$$\left(\frac{A_c^{ij} + A_c^{ji}}{2}\right)^T P + P\left(\frac{A_c^{ij} + A_c^{ji}}{2}\right) < 0 \tag{35}$$

for $i < j \leq L$, where A_{ii}^c denotes the state space matrix corresponding to feedback connection of local model in i'th rule and regulator designed for the same rule. A_{ij}^c denotes the state space matrix corresponding to feedback connection of local model in i'th rule and regulator designed for the j'th rule. In the same way for A_{ji}^c (Ma et al., 1998). These inequalities can be efficiently solved numerically through a linear matrix inequality (LMI) (Boyd et al., 1994) framework.

However, second condition expressed by (35) is very conservative and requires that feedback connection of every regulator with every local model in rule-base to be stable. According to 2-D triangular membership functions used in antecedent part, as shown in Figure 6, it is sufficient to check second condition only for adjacent membership functions. Because, for triangular memberships functions, with a specified value of premise vector (u_{pc}, w_{gc}), only different regulator from other rules that are corresponding to adjacent rules will be excited and membership degrees of remaining membership functions will be equal to zeros.

All 81 state-space models of closed-loop systems are of order 6. They are used to construct a polytopic system and LMI (Boyd et al., 1994) is used to find desired positive definite matrix as below:

$$P = \begin{bmatrix} 4.50 & 8.64 & -0.12 & -5.52 & -0.53 & -0.34 \\ 8.64 & 59.06 & 0.47 & 3.18 & -41.31 & 6.80 \\ -0.12 & 0.47 & 0.87 & 0.56 & 0.18 & -0.36 \\ -5.52 & 3.18 & 0.56 & 21.7 & 9.03 & 8.61 \\ -0.53 & -41.3 & 0.18 & 9.03 & 52.2 & -0.51 \\ -0.34 & 6.80 & -0.36 & 8.61 & -0.51 & 17.1 \end{bmatrix} \quad (36)$$

SIMULATION STUDY

OLGA® is a very realistic simulator for simulation of multi-phase flows in pipeline. For our purpose, network set up shown in Figure 9 is used. Four nodes are used in this scheme. “Gas inlet” and “Oil inlet” are closed nodes and “well Head” is a constant pressure node, representing separator pressure. “Injection” is a merge node that connects “annulus” and “bottom hole” branches to “tubing” branch. “Bottom hole” is a 100 meters vertical pipe that brings liquid form well to injection point. Tubing is a 2400 meters vertical regular pipe, and annulus is a 2400 meter annular flow pipe. “Source” is a constant flow rate source of injection gas. To ensure that liquid don’t come to annulus a “check valve” is used, that is coincide with injection orifice. Another check valve is used at well head. “Production Choke” is a choke, its opening portion is dictated with control signal coming from fuzzy LQG regulator that is implemented in MATLAB and connected with OLGA® via Matlab-Olga-Link® Toolbox.

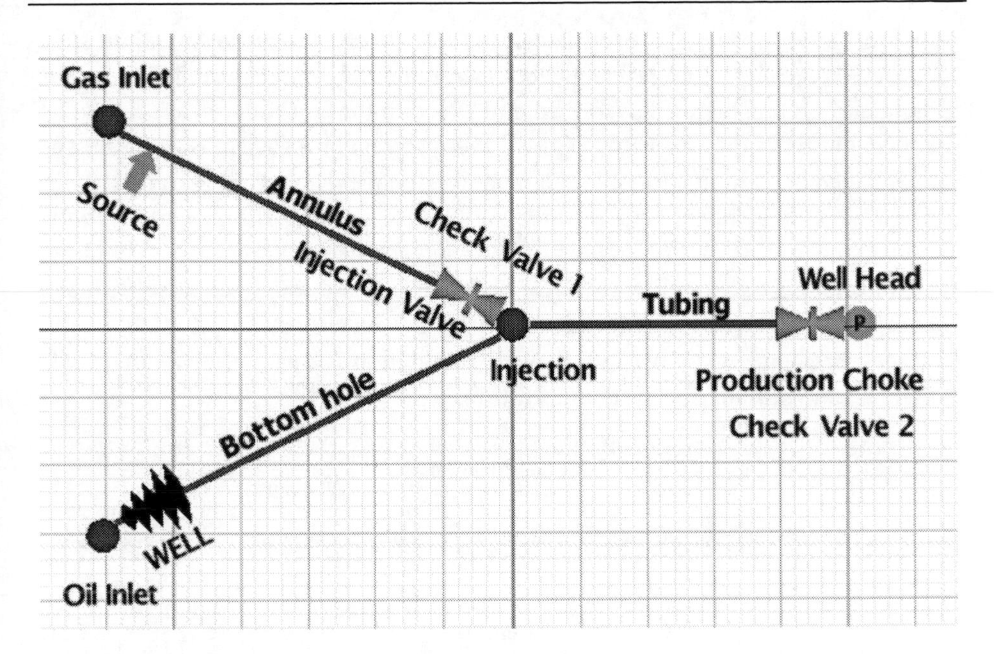

Figure 9. Schematic of network used for simulation of model in OLGA®v5.0.

For implementation of fuzzy regulator in MATLAB® environment, a Gaussian noise with standard deviation of 0.01 (process noise) was added to w_{gc} and similarly Gaussian measurement noises with standard deviations of 10 and 10,000 were added to $y_1(t)$ and $y_2(t)$, respectively. As a consequence, the covariance matrices would be:

$$Q = E(ww^T) = 1\times 10^{-4} \tag{37}$$

$$R = E(vv^T) = \begin{bmatrix} 10^2 & 10^3 \\ 10^3 & 10^8 \end{bmatrix} \tag{38}$$

$$N = E(wv^T) = \begin{bmatrix} -10^{-4} & 1 \end{bmatrix} \tag{39}$$

N value shows that measurement and process noises are fairly uncorrelated. Using the model bank derived from linearization of nonlinear model in different operating points, as described before, and the above uncertainty levels, a Kalman estimator is designed for each local model. Then, the corresponding state-feedback controller gain is calculated using the LQ technique. The resulting individual LQG regulators are formed and saved in the controller bank. Using 2-D membership functions shown in Figure 6 the fuzzy control system is constructed.

For demonstration, variable command signals, shown in Figure 10, are introduced to the model implemented in OLGA®. Using the proposed control strategy, the oscillatory behaviour of system, shown in Figure 2, is stabilized which is demonstrated in Figure 11 for both the real and estimated state variables. Figure 10, also shows the required control signal which has an affordable value. The measurements and flow rate of well production are shown in Figure 12 for both open and closed-loop system operations.

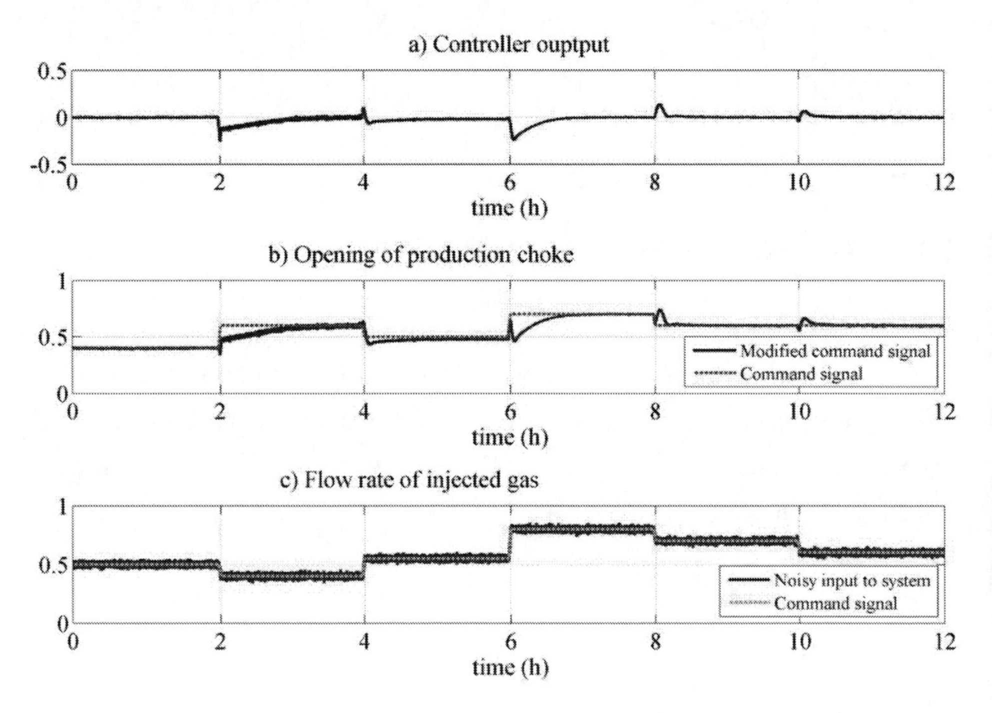

Figure 10. Controller output signal and system inputs, a) Controller output, b) Opening of production choke, c) Flow rate of injected gas.

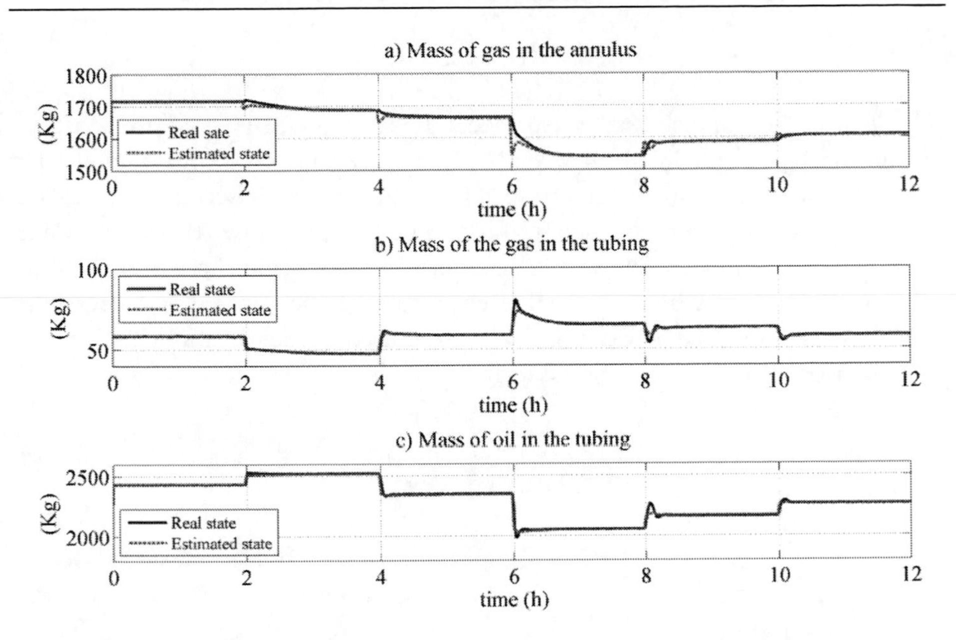

Figure 11. State variables (solid) and their estimates (dashed) for closed loop system, a) Mass of gas in the annulus, b) Mass of the gas in the tubing, c) Mass of the oil in the tubing.

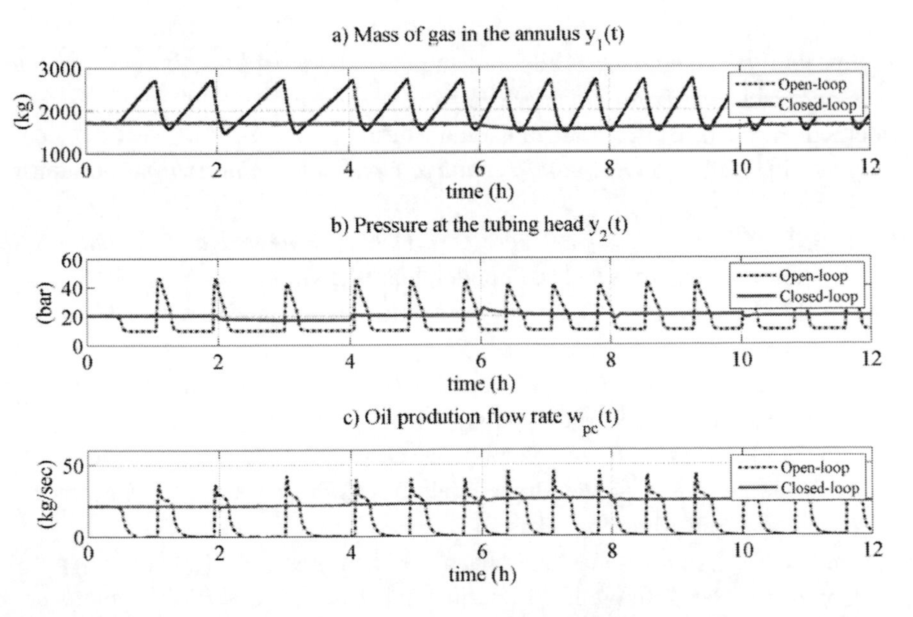

Figure 12. $y_1(t)$, y_2(t) and $w_{pc}(t)$ for open-loop and closed-loop conditions, a) Mass of gas in the annulus, b) Tubing head pressure, c) Production flow rate.

Conclusions

State-space stability analysis of local linear models is used to construct stability maps for gas-lifted oil wells. On the other hand, a TS fuzzy model for highly nonlinear system dynamics of gas-lift is constructed based on the proposed stability map. Based on the resulting fuzzy model, fuzzy observer and controller successfully are utilized for state estimation and stabilization in a realistic case study of unstable gas-lifted oil well. Hence, fuzzy logic concept is successfully used in creating a link between stability maps used by petroleum engineers and model-based estimation and control techniques.

References

Aamo, O. M., Eikrem, G. O., Siahaan, H. and Foss, B. (2004), "Observer Design for Gas Lifted Oil Wells," *American Control Conference,* Boston, Massachusetts.

Aamo, O. M., Eikrem, G. O., Siahaan, H. and Foss, B. (2005), "Observer design for multiphase flow in vertical pipes with gas-lift - theory and experiments," *Journal of Process Control,* 15, 247–257.

Alhanati, F. J. S., Doty, D. R. and Lagerief, D. D. (1993), "Continuous Gas-Lift Instability: Diagnosis, Criteria, and Solutions. *68th Annual Technlcel Conference and Exhibition of SPE,* Houston, Texas.

Bloemen, H. H. J., Belfroid, S. P. C. and Sturm, W. L. (2004), "Soft Sensing for Gas-Lift Wells," *SPE Annual Technical Conference and Exhibition,* Houston, Texas.

Boyd, S. P., Ghaoui, L. E. and Feron, E. (1994), *Linear Matrix Inequalities in Systems and Control Theory,* Philadelphia: SIAM.

Brown, K. E. (1982), "Overview of Artificial Lift Systems," *Journal of Production Technology, SPE no. 1979.*

Burl, J. B. (1999), *Linear Optimal Control: H-2 and H-inf Methods,* Menlo Park, California: Addison Wesley Longman Inc.

Dalsmo, M., Halvorsen, E. and Slupphaug, O. (2002), "Active Feedback Control of Unstable Wells at the Brage Field," *Annual Technical Conference and Exhibition,* San Antonio, Texas.

Eikrem, G. O., Aamo, O. M., Siahaan, H. and Foss, B. (2004a), "Anti-Slug Control of Gas-Lift Wells - Experimental Results," *6th IFAC Symposium on Nonlinear Control Systems,* Stuttgart, Germany.

Eikrem, G. O., Foss, B., Imsland, L., Hu, B. and Golan, M. (2002), "Stabilization of Gas Lifted Wells," *15th IFAC World Congress,* Barcelona, Spain.

Eikrem, G. O., Imsland, L. s. and Foss, B. (2004b), "Stabilization of Gas Lifted Wells Based on State Estimation," *International Symposium on Advanced Control of Chemical Processes,* Hong Kong, China.

Fairuzov, Y. V., Guerrero-Sarabia, I., Calva-Morales, C., (2004), "Stability Maps for Continuous Gas-Lift Wells: A New Approach to Solving an Old Problem," *SPE Annual Technical Conference and Exhibition,* Houston, Texas.

Hu, B. and Golan, M. (2003), "Gas-lift instability resulted production loss and its remedy by feedback control: dynamical simulation results," *SPE International Improved Oil Recovery Conference in Asia Pacific,* Kuala Lumpur, Malaysia.

Imsland, L., Foss, B. A. and Eikrem, G. O. (2003), "State Feedback Control of A Class of Positive Systems: Application To Ggas Lift Stabilization," *7th European Control Conference,* Cambridge, UK.

Jansen, B., Dalsmo, M., Nøkleberg, L., Havre, K., Kristiansen, V. and Lemetayer, P. (1999), "Automatic Control of Unstable Gas Lifted Wells," *SPE Annual Technical Conference and Exhibition,* Houston, Texas.

Layne, J. R. and Passino, K. M. (1996), "Fuzzy Dynamic Model Based State Estimator," *IEEE international Symposium on Intelligent Control,* Dearborn, MI.

Ma, X.-J., Sun, Z.-Q. and He, Y.-Y. (1998), "Analysis and Design of Fuzzy Controller and Fuzzy Observer," *IEEE Transaction on Fuzzy Systems,* 6.

Passino, K. M. and Yurkovich, S. (1998), *Fuzzy Control,* Menlo Park, CA: Addison Wesley Longman, Inc.

Poblano, E., Camacho, R. and Fairuzov, Y. V. (2005), "Stability Analysis of Continuous-Flow Gas Lift Wells," *SPE Production and Facilities Journal.*

Scandpower (2006), *OLGA Verion 5 User's Manual*: Scandpower.

Simon, D. (2003), "Kalman filtering for fuzzy discrete time dynamic systems," *Applied Soft Computing,* 3, 191–207.

Simon, D. (2006), *Optimal State Estimation, Kalman, H-inf and Nonlinear Approches,* Hoboken, New Jersey: Wiley-Interscience.

Sinegre, L. (2006), "Etude des instabilités dans les puits activés par gas-lift," Ph.D. dissertation, Spécialité "Mathématiques et Automatique", Ecole des Mines de Paris, Paris.

Sinegre, L., Petit, N. and Menegatti, P. (2006), "Predicting instabilities in gas-lifted wells simulation," *2006 American Control Conference.* pp. 5530-5537, Minneapolis, Minnesota.

Sonntag, R. E., Borgnakke, C. and Wylen, G. J. V. (2003), *Fundamentals of Thermodynamics,* New York: John Wiley and Sons.

Terre, A. J., Schmidt, Z., Blais, R. N., Doty, D. R. and Brill, J. P. (1987), "Casing Heading in Flowing Oil Wells," *SPE Production Engineering, SPE no. 13801.*

In: Natural Gas Systems
Editor: Rafiq Islam
ISBN: 978-1-61324-158-5

Chapter 3

USING REBOILER DUTY OF THE REGENERATOR COLUMN AS A MEASURE OF THE ECONOMIC FEASIBILITY OF SOUR GAS TREATING: A CARIBBEAN REFINERY AS A STUDY CASE

M. Steven[1,2], *S. Sinanan*[1] *and C. Riverol*[2,*]
[1]Chemical Engineering Department, University of West Indies, Trinidad
[2]Petrotrin, Point a Pierre, Trinidad, Republic of Trinidad and Tobago

ABSTRACT

Several papers describe as sour gases should be treated but the energy consumption in the process is a problem is being encountered. The energy consumption can be affected for the type of amine used in the absorption process (the reboiler can need more or less steam). Presently at Petrotrin, the results shown in this article indicated that the MDEA shows a lower reboiler duty for regeneration, such that the use of Methyldiethanolamine, MDEA is economically feasible and as such, it is recommended that the existing amine Diethanolamine or other amine mixtures system be replaced by the use of Methyldiethanolamine, MDEA.

* Corresponding author: criverol@eng.uwi.tt, Telephone: (1868)6622002, Fax (1868)6624414

Keywords: Reboiler, CO_2 absorption, selectivity, sour gas, steam, heat, economic.

1. INTRODUCTION

The removal of hydrogen sulfide (H_2S) from liquid and gaseous hydrocarbon streams has long been desired to reduce the malodorous and corrosive characteristics of streams rich in this compound [Salvage, 1981]. This gas is extremely poisonous and very harmful to humans. More recently, governmental legislation requiring reduced sulfur pollutant emissions to the atmosphere from various fuels is becoming more prevalent throughout the world. To date, the most economical and widely used process to efficiently remove H_2S and other acidic contaminants is a continuous absorption/regeneration process employing an aqueous solution of basic alkanol amine. The alkanol amines most commonly used are Monoethanolamine (MEA): $HOCH_2CH_2NH_2$ or Diethanolamine (DEA): $(HOCH_2CH_2)_2NH$. These ethanol amines have high affinity for H_2S and low solubility in hydrocarbons and are also generally applicable to CO_2 removal, see [Astarita, 1982]. At Petrotrin's Pointe-a-Pierre refinery, sour gas is treated for H_2S removal by four absorbers, utilizing a solution of 10% by weight Diethanolamine solution. The rich amine or amine rich in H_2S from all the absorbers is routed to the Regenerator Column at No. 2 Girbotol where the H_2S is stripped off the amine. The H_2S that is removed from the rich amine is routed to the Cyanamid Acid Recovery Plant (CARP) or to the Sulfur Recovery Unit (SRU), where the H_2S is converted to Sulphuric Acid or Sulfur respectively. A simplified process flow diagram can be seen in Figure 1.

Due to the high carbon dioxide, CO_2 content of some of the sour gas streams in the refinery [Salvage, 1981], DEA absorbs the CO_2 as well as the H_2S which in turn reduces the H_2S purity of the H_2S rich Regenerator offgas. Low H_2S purity adversely affects the operations of the CARP.

Literature indicate that Methyldiethanolamine, MDEA has several benefits over the use of conventional amine MEA or DEA such as higher selectivity towards H_2S and lower energy requirements for regeneration [Salvage, 1981]. Thus the result of this research can be used as preliminary findings, which will determine whether switching to MDEA may lead to high H_2S purity in the regenerator offgas and lower reboiler duty leading to reduced steam usage thus determining the economic feasibility of switching to MDEA.

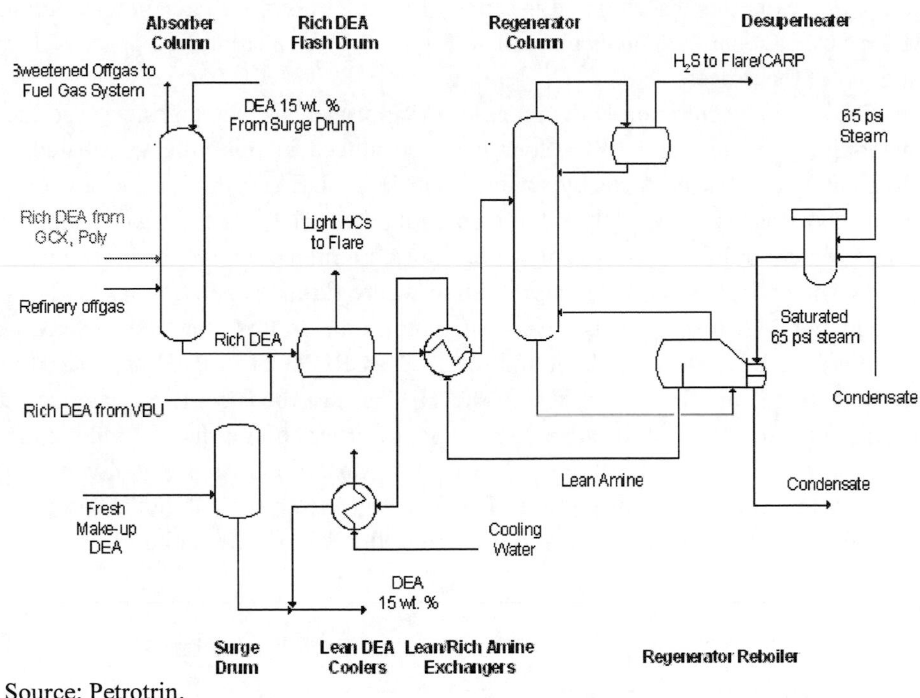

Source: Petrotrin.

Figure 1. Simplified Process Flow Diagram of the Girbotol Unit.

2. Process Description

At Petrotrin's Pointe-a-Pierre Refinery, sour gas is treated for H_2S removal. Currently, there are four absorbers at the No. 2 Girbotol Unit, Visbreaker Unit, Gas Concentration Unit and the Catalytic Polymerisation Unit which utilise Diethanolamine, DEA solution for H_2S removal. The Catalytic Polymerisation Unit is used to reduce propene/propane inventory by converting it to poly gasoline (high octane gasoline). The unit removes H_2S and mercaptan sulfur from the propene/propane produced as well as reduces the unsaturated content of the propene/propane leaving the unit, thereby improving the quality of LPG product. The feed is purified in the treatment section at a capacity of 3000 BPD. Feed purification removes H_2S and mercaptan sulphur from the charge to meet the corrosion requirements of LPG. Fresh propene/propane is contacted with lean DEA in the amine scrubber which contains approximately 2500 ppm of H_2S. The rich amine exits the absorber and is sent to No. 2 Girbotol's Regenerator column where further separation takes place. The Figure 2 gives a simplified process flow diagram of the process.

The Gas Concentration Unit processes compressed gas, compression gasoline, unstabilised gasoline and LPG. Sour gases produced in this unit are treated at 1300 BPD for H_2S removal by contact with lean DEA in the amine absorber which contains approximately 22500 ppm of H_2S. The sweet offgas from the column enters the fuel gas system while the rich amine exits the absorber and is sent to No. 2 Girbotol's Regenerator column where further separation takes place.

The No.2 Girbotol Unit is a gas sweetening unit. In this unit, H_2S rich gases from the No. 1 Hydrotreater Unit and the No. 2 CRU at 1150 BPD are routed to the Girbotol Amine Absorber, where the H_2S is absorbed from the gases by a solution of circa 10% wt DEA solution. The absorber column has 19 bubble cap trays and operates at 620.528kPa (90 psig). The absorption of H_2S is favored by high pressures and low temperatures. The sweet gas from the column is routed to the fuel gas system and the rich amine is sent to the Regenerator column.

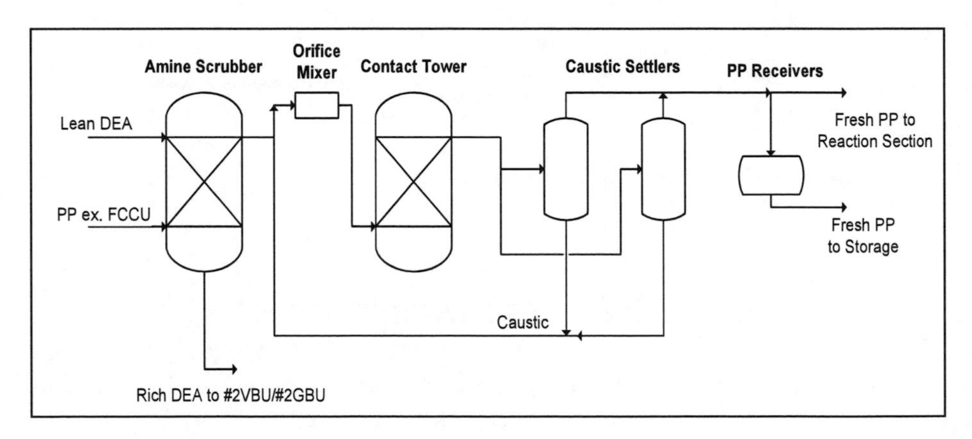

Figure 2. Process Flow Diagram of the Catalytic Polymerisation Unit Treatment Section. Source: Petrotrin.

The Cyanamid Acid Recovery Plant requires high purity H_2S and as a result of the low purity presently being produced by the Girbotol and Visbreaker Units, the CARP uses H_2S from another plant, No. 2 HTU and as such, Girbotol has to flare the H_2S being produces resulting in SO_2 emission to the atmosphere.

Visbreaking is a mild form of thermal cracking which significantly lowers the viscosity of heavy crude oil residue. The visbreaking unit is thus designed to process vacuum reside of an indigenous Trinidad Crude. The vapour product exiting the overhead of the unit's fractionator is routed to the amine absorber, where 1050 BPD of sour offgas is treated. In the amine absorber, H_2S is removed from the sour offgas by contact with a lean amine solution. The H_2S rich amine is routed to the battery limits for regeneration. The Visbreaker sweetened offgas is primarily used as fuel in the Visbreaker heater and excess is exported to the fuel gas system.

The rich amine solution exiting all four absorbers is mixed and routed to the Rich Amine Flash Drum which operates at about 34.473 kPa (5 psig). This vessel is designed to flash off light hydrocarbons and also skim off heavy hydrocarbon liquid from the amine solution. From the rich amine flash drum, the amine solution is pumped through the rich/lean amine heat exchanger, where the amine solution is heated to about 98.89^0C (210^0F) before entering the Regenerator Column. In the regenerator column, the H_2S is stripped off the amine solution. This column has 19 float valve trays and operates at 82.737-110.316 kPa (12-16 psig). This column has a 448.159 kPa (65 psig) steam desuperheated kettle reboiler and a water cooled overhead condenser. The regenerator column is operated to achieve <50 grains H_2S/USgal solution. From the kettle reboil, the lean amine solution is cooled and routed to the onsite absorber and also the absorbers at the Catalytic Polymerisation Unit, Visbreaker Unit and Gas Concentration Unit which remove the H_2S from the propene/propane streams, Visbreaker offgas and lean gas respectively. From the top of the regenerator column, H_2S is taken off and sent to the Cyanamid Acid Recovery Plant and the Sulphur Recovery Unit where sulphuric acid and sulphur are produced respectively.

3. Results and Discussion

The results of this project are based on the HYSYS simulations which were simulated as a model of the existing system at Petrotrin , see [Luyben (2002) and Hyprotech (1999) manual]. The following results were obtained by using three different solvents for absorption of H_2S from the sour gas streams namely, Diethanolamine DEA, Monoethanolamine MEA and Methyldiethanolamine, MDEA.

The steam required for regeneration has been calculated and a summary is shown in Table 1. This table shows that the cost of regeneration using MEA is greater than that by using DEA, which is greater than that by using MDEA. It costs US$16.086 an hour to use steam for regeneration by using MDEA and US$20.02 an hour by using DEA and US$33.30 an hour by using MEA.

Table 1. Flowrates of CO_2 in the top and bottom stream of the amine absorbers using different amines

	Gas Concentration Unit		Visbreaker Unit	
	Top Stream flowrate (kg/h)	Bottom Stream flowrate (kg/h)	Top Stream flowrate (kg/h)	Bottom Stream flowrate (kg/h)
MEA	33.03	54.162	58.28	127.09
DEA	64.4	28.196	113.38	66.16
MDEA	91.69	6.0	159.89	14.07

Table 2. The compositions of the inlet and outlet streams of the Regenerator Column by using Diethanolamine, DEA

	Feed To Regenerator Column	H_2S To CARP and SRU	Hot Lean Amine Exit Regen Column
DEA	0.0186	0.0000	0.0187
H_2O	0.9751	0.0452	0.9807
H_2S	0.0054	0.8050	0.0006
CO_2	0.0009	0.1450	0.0000
H_2	0.0000	0.0010	0.0000
Methane	0.0000	0.0012	0.0000
Ethane	0.0000	0.0006	0.0000
Ethylene	0.0000	0.0003	0.0000
Propane	0.0000	0.0003	0.0000
Propene	0.0000	0.0014	0.0000
i-Butane	0.0000	0.0000	0.0000
n-Butane	0.0000	0.0000	0.0000
N_2	0.0000	0.0000	0.0000

Therefore. it is more economical to use MDEA rather than MEA or DEA for steam generation. The price of MDEA presently is US$1.00 which is considerably less than the price of MDEA in the past and as such it can be even more economical to use MDEA for the selective absorption of H_2S than any of the other mentioned amines in the past.

From the above analysis, it can be seen that the use of MDEA has its advantage in all aspects of sour gas treating, from absorption, to regeneration to economic. The Cyanamid Acid Recovery Plant requires that the offgas from the regenerator column be very pure in H_2S, so that its operation is carried out efficiently. That is, contains very little or no CO_2. From the Table 2 until 4 the composition of the inlet and outlet streams of the regenerator column is given utilizing all three amines for comparison.

The reboiler duty of the regenerator column will be used as a measure of the economic feasibility of switching from DEA to MDEA by determining the heat/steam requirement. The Table 5 shows the reboiler duty of each of the regenerator columns using MEA, DEA and MDEA. In order to determine the heating requirement (steam) for the regenerator column, the following calculation is performed:

$$\text{Mass Flow Rate Of Steam [kg/h]} = \frac{\text{Reboiler Duty [kJ/h]}}{\text{Latent Heat of Vaporisation [kJ/kg]}} \quad (1)$$

The latent heat of vaporization was determined from the steam tables at 45 psig 300°F = 2130.3 kJ/kg

Also, the cost of steam at 65psig is = US$1.51 / 1000 lb

Since 1 lb = 0.4536 kg

Cost of steam in kg = US$1.51 / 453.6 kg

Therefore the cost of steam can be calculated and are show in Table 6.

Moreover, by using MDEA, the amount of steam required for regeneration was considerably less. In fact, when a flow of 11000 b/d was used, the cost of steam was US$16.09 per hour but when the feed was cut to 5200 b/d a cost of US$7.57 per hour was required. This is approximately 50% of the initial cost required for regeneration. The above analysis clearly demonstrates that MDEA has many versatile features that make it attractive for selective H_2S removal which includes a large decrease in the cost of steam as well as MDEA.

Table 3. The compositions of the inlet and outlet streams of the Regenerator Column by using Monoethanolamine, MEA

	Feed To Regenerator Column	H_2S To CARP and SRU	Hot Lean Amine Exit Regen Column
MEA	0.0316	0.0000	0.0317
H_2O	0.9613	0.0452	0.9664
H_2S	0.0054	0.7971	0.0010
CO_2	0.0017	0.1525	0.0009
H_2	0.0000	0.0010	0.0000
Methane	0.0000	0.0013	0.0000
Ethane	0.0000	0.0006	0.0000
Ethylene	0.0000	0.0003	0.0000
Propane	0.0000	0.0004	0.0000
Propene	0.0000	0.0015	0.0000
i-Butane	0.0000	0.0000	0.0000
n-Butane	0.0000	0.0000	0.0000
N_2	0.0000	0.0000	0.0000

Table 4. The compositions of the inlet and outlet streams of the Regenerator Column by using Methyldiethanolamine, MDEA

	Feed To Regenerator Column	H_2S To CARP and SRU	Hot Lean Amine Exit Regen Column
MDEA	0.0165	0.0000	0.0165
H_2O	0.9779	0.0452	0.9829
H_2S	0.0054	0.9203	0.0005
CO_2	0.0002	0.0290	0.0000
H_2	0.0000	0.0011	0.0000
Methane	0.0000	0.0013	0.0000
Ethane	0.0000	0.0007	0.0000
Ethylene	0.0000	0.0004	0.0000
Propane	0.0000	0.0004	0.0000
Propene	0.0000	0.0016	0.0000
i-Butane	0.0000	0.0000	0.0000
n-Butane	0.0000	0.0000	0.0000
N_2	0.0000	0.0000	0.0000

Table 5. The Reboiler Duty of the Regenerator Column using different solvents

Reboiler Duty		
	Heat Flow Btu/hr	Heat Flow kJ/hr
MEA	2.02×10^7	2.128×10^7
DEA	1.21×10^7	1.279×10^7
MDEA	9.75×10^6	1.028×10^7

Table 6. The cost of steam required for regeneration

	Heat Flow kJ/h	Mass Flowrate of steam kg/h	Cost of steam US$ /h
MEA	2.128×10^7	9989.2	33.30
DEA	1.279×10^7	6003.8	20.02
MDEA	1.028×10^7	4825.6	16.08

CONCLUSION

The objectives of this research have been met and the proposition made has been validated by conclusive results. Using MEA, which has a low acid gas loading, requires a large amount of steam for regeneration and therefore adds cost to the plant. Also, there are problems of corrosion to be considered when using MEA.

Methyldiethanolamine, MDEA was investigated as a suitable alternative to DEA and from the results of this analysis; the cost of heating requirement needed for regeneration is low compared to both MEA and DEA and therefore cost is reduced. Also, the cost of MDEA has recently dropped as discussed and will therefore mitigate cost. The properties of MDEA are very unique and because of this, it has been found to be a suitable replacement for the DEA system.

References

Astarita, G. 1995 Gas Treating with Chemical Solvents. University of Naples and University of Delaware; David W. Savage, Exxon Research and Engineering Company; Attilio Bisio, Exxon Research and Engineering Company. John Wiley and Sons.

Hyprotech Limited Refinery Simulation Workshop – November 1999.

Luyben W. 2002 Plantwide Simulators in Chemical Processing and Control. Marcel Dekker, 1st edition, New York.

Savage, D. W., and E. W. Funk,. 1981 "Selective absorption of H2S and CO2 into Aqueous Solutions of MDEA," paper presented at AIChE, Houston Texas, April 5-9.

In: Natural Gas Systems
Editor: Rafiq Islam

ISBN: 978-1-61324-158-5

Chapter 4

FLAMMABILITY AND INDIVIDUAL RISK ASSESSMENT FOR NATURAL GAS PIPELINES

M. Enamul Hossain, Chefi Ketata, M. Ibrahim Khan and M. Rafiqul Islam*
Department of Civil and Resource Engineering,
Dalhousie University, Halifax, Canada

ABSTRACT

Natural gas and oil are mainly supplied and transmitted through pipelines. The safety and risk factors for transporting natural gas through pipelines is an important issue. This paper develops a comprehensive model for the individual risk assessment for natural gas pipelines. Presently available models related to pipeline risk assessment are also examined and their shortcomings identified. To overcome these limitations, a new concept of individual risk is introduced. It combines the flammability limit with existing individual risk for an accidental scenario. The new model determines the major accidental area within a locality surrounded by pipelines. Finally, the proposed model is validated using field data. This innovative model applies to any natural gas pipeline risk assessment scenario.

Keywords: natural gas pipelines, flammability limit, individual risk, explosion hazard.

* Corresponding author: Email: rafiqul.islam@dal.ca

1. INTRODUCTION

Natural gas is one of the most widely used domestic fuels in industrialized countries. The consumption of natural gas is continuously increasing. As a result, complex piping systems are being installed to transport and distribute the gas for end users. These pipeline networks are mostly installed in urban zones, i.e. in highly populated areas. Therefore, accidental gas releases can cause significant environmental damages, economic losses and injury to the population (Khan et al., 2006). Moreover, gas piping systems are mostly installed at underground. They are often damaged by various activities. It is reported that approximately 67% of accidents involving natural gas occur in piping systems (Arnaldos et al., 1998).

The failure of natural gas pipelines may occur due to natural or man-made disasters such as earthquake, hurricane, sabotage, overpressure, flood, corrosion, or fatigue failures. The failure rate is also influenced by design factors, construction conditions, maintenance policy, technology usage and environmental factors. All kinds of accidents in pipelines are determined by the risk assessment and management (Ramanathan, 2001). Risk assessment is the process of obtaining a quantitative estimate of a risk by evaluating its probability and consequences. Risk is generally referred to the potential for human harm. This risk represents a hazardous scenario, which is a physical or societal situation. If encountered, it could initiate a range of undesirable consequences. The most frequent cause is perforation of the pipe or complete fracture. Gas will be released to the environment at a flow rate depending on the hole diameter and the pressure in the pipe until the release is stopped automatically by means of a regulator, as a reaction to excessive flow rate, or manually.

Natural gas pipelines failures are potentially hazardous events especially in urban areas and near roads. Therefore people around the pipeline routes are subject to significant risk from pipeline failure. The hazard distance associated with the pipeline ranges from under 20 m for a smaller pipeline at lower pressure, up to over 300 m for a larger one at higher pressure (Jo and Ahn, 2002). So it is essential to study the level of pipeline safety for a better risk assessment and management.

To determine the individual risk of an explosion hazard, flammability limits data are essential in a natural gas pipeline. Flammability limits are commonly used indices to represent the flammability characteristics of gases. These limits can be defined as those fuel-air ratios within which flame propagation can be possible and beyond which flames cannot propagate. By definition there are two flammability limits namely lower flammability limit (LFL) and upper flammability limit (UFL). LFL can be defined as the leanest fuel limit up to which the flame can propagate and the richest limit is called as UFL (Liao et al., 2005). The flammability limit criterion, and other related parameters have been broadly discussed in the available literature (Vanderstraeten et al., 1997; Kenneth et al., 2000; Kevin et al., 2000; Pfahi et al., 2000; Wierzba and Ale, 2000; Mishra and Rahman, 2003; Takahashi et al., 2003). The objectives of this study are to predict a future outcome with certainty and to eliminate future risk. This study introduces a new dimension of risk assessment combining risk due to flammability limit and lethal failure in the accident scenarios.

2. Pipeline Risk Management

Natural gas pipelines are an elongated pressure container with unlimited flow. They transport large quantities of natural gas at elevated pressures. The pipelines represent a hazardous risk to nearby population and facilities, in addition to business interruption concerns. Although underground burial of pipelines is recommended, it does not prevent pipeline accidents from happening, since gas leakage and pipeline failure are still possible. A means for emergency isolation should be supplied at pipeline entries and exits from various facilities. For integrity assurances, pipelines should be verified regularly for failures and leakages at vulnerable locations including weld joints and flange connections. These are usually checked using testing techniques, such as ultrasound, x-ray, and die penetrants.

The primary factor affecting pipeline hazardous incidents in normal situations is corrosion. Therefore, it is important to take care of the pipelines by using proper anti-corrosion materials. Furthermore, pipeline failure can result from third party activity, sabotage, or natural disasters. Figure 1 illustrates the risk management approach for natural gas pipelines comprising the following steps:

1. Piping system identification;
2. Operations information;
3. Risk assessment;
4. Strategy;
5. Actions;
6. Evaluation.

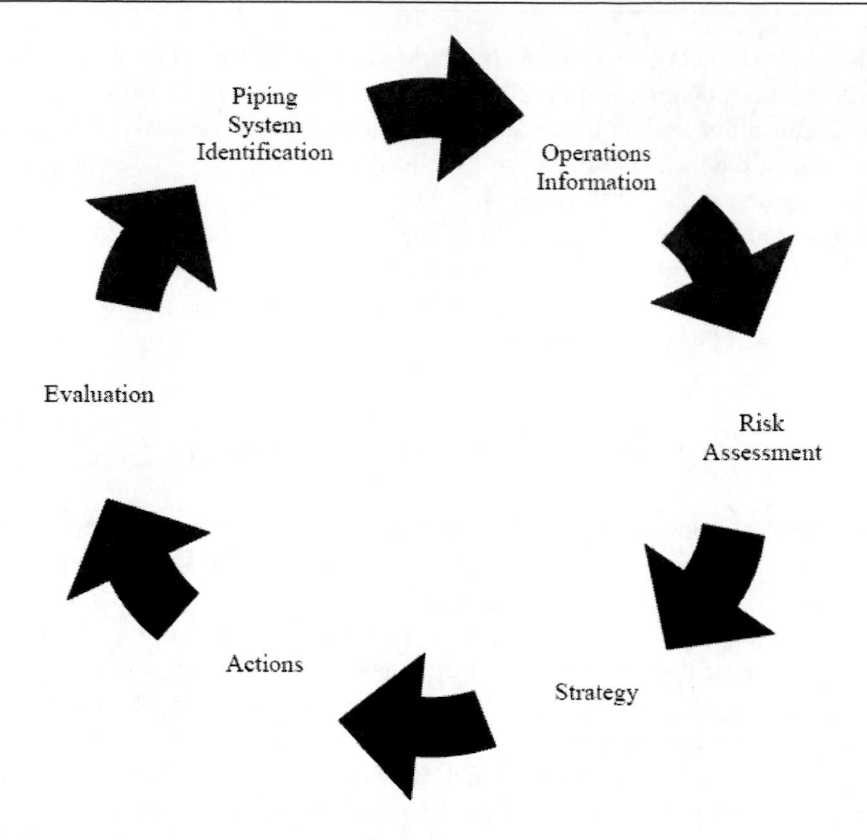

Figure 1. Risk management for natural gas pipelines.

3. Risk Assessment

In order to assess the risk regarding the natural gas pipeline, it is necessary to evaluate probable undesirable consequences resulting from any pipeline leakage or rupture.

The quantitative risk can be estimated due to flammability limit for a natural gas pipeline. Risk has been described as individual risk, societal risk, maximum individual risk, average individual risk of exposed population, average individual risk of total population, and average rate of death (TNO Purple Book, 1999; Jo and Ahn, 2002 and 2005). This study will demonstrate how the individual risk is influenced by flammability limits and other related parameters.

The failure rate of pipelines depends on various parameters such as soil conditions, coating type and properties, design considerations, and pipeline age.

So a long pipeline is divided into sections due to significant changes of these parameters. Considering a constant failure rate, the individual risk (Jo and Ahn, 2005) can be written as:

$$IR = \sum_i \varphi_i \int_{l_-}^{l_+} p_i dl \quad (1)$$

where

φ_i = Failure rate per unit length of the pipeline associated with the accident scenario i due to soil condition, coating, design and age, *1/year km*

l = Pipeline length, *m*

p_i= Lethality associated with the accident scenario i

$l_\pm$ = Ends of the interacting section of the pipeline in which an accident poses hazard to the specified location, *m*

The release of gas through a hole on the pipeline causes explosion and fire in the natural gas pipeline and the surrounding area, which creates accidents. The effected section causes a hazard distance. The release rate of natural gas and hazard distance are correlated (Jo and Ahn, 2002):

$$r_h = 10.285\sqrt{Q_{eff}} \quad (2)$$

where

Q_{eff} = Effective release rate from a hole on a pipeline carrying natural gas, *kg/sec*

r_h = Hazard distance, *m*

The hazard distance is the distance within which there is more than one percent chance of fatality due to the radiational heat of jet fire from pipeline rupture. Figure 2 shows the geometric relations among the variables in specified location from a natural gas pipeline. From this figure, the interacting section of a straight pipeline, h, from a specified location, is estimated by the following equation (Jo and Ahn, 2005):

$$l_\pm = \sqrt{106Q_{eff} - h^2} \quad (3)$$

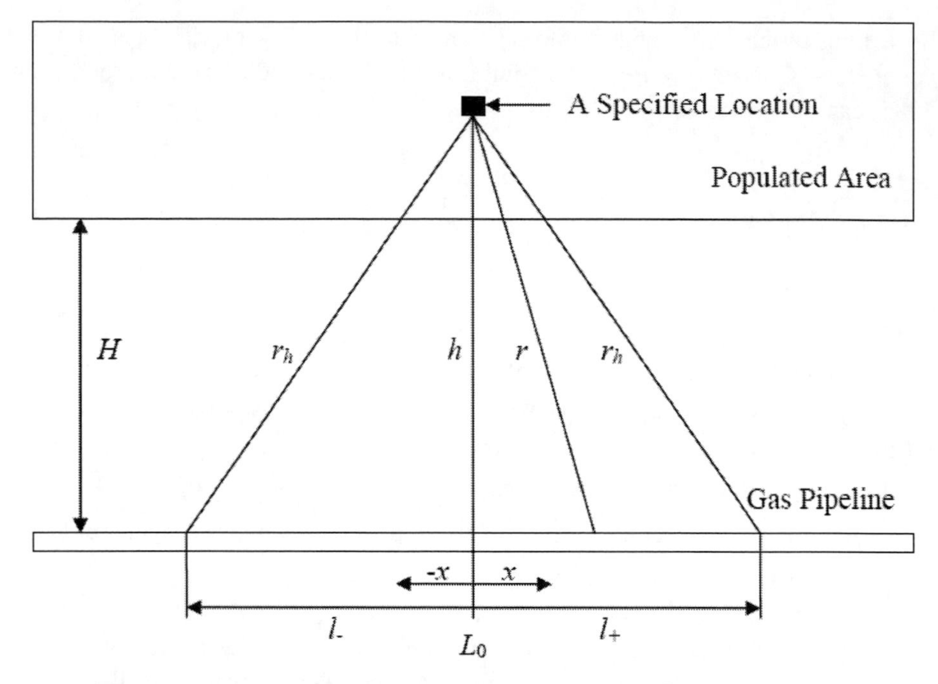

Figure 2. Individual risk variables (Redrawn from Jo and Ahn, 2005).

Jo and Ahn (2005) show the different causes of failure based on hole size and other activities. The external interference by third party activity is the major cause of key accidents related to hole size. Therefore, a more detailed concept is extremely required to analyze the external interference. The third party activity on pipeline depends on several factors, such as pipe diameter, cover depth, wall thickness, population density, and prevention method. The failure rate of a pipeline can be estimated by some researchers (Jo and Ahn, 2005; John et al., 2001).

4. Effects of Composition on Flammability Limit

An experimental study is usually conducted to investigate the effects of concentration or dilution in natural gas – air mixture by adding CO_2, N_2 gas. The limit ranges are 85-90% of N_2 and 15-10% of CO_2 by volume. This is quite practical considering natural gas stoichiometric combustion at ambient temperature. Flammability experiments have been performed to simulate real explosions in order to prevent hazards in the practical applications (Liao et al.,

2005). Table 1 shows the flammability limit data for methane-air and natural gas-air flames according to Liao et al. (2005).

LFL depends on the composition of fuel mixture in air. This value can be estimated by LeChatelier's rule (Liao et al., 2005):

$$LFL = \frac{100}{\sum (C_i / LFL_i)} \quad (4)$$

where
LFL = Lower flammability limit of mixture (vol. %)
C_i = Concentration of component i in the gas mixture on an air-free basis (vol. %)
LFL_i = Lower flammability limit for component i (vol. %).

Table 1. Flammability limit data (vol %) for methane-air and natural gas-air flames (quiescent mixtures with spark ignition)

Mixture	Test Condition	LFL (vol %)	UFL (vol %)
NG-air	1.57 L chamber	5.0	15.6
	LeChatelier's rule	4.98	-
Methane-air	8 L chamber	5.0	-
	20 L chamber	4.9	15.9
	120 L chamber	5.0	15.7
	25.5 m^3 sphere	4.9, 5.1 $\pm$ 0.1	-
	Flammability tube	4.9	15.0

The estimation of LeChatelier's rule is shown in Table 1 and is plotted in Figure 3 as well. The reliance of natural gas flammability limit upon ethane concentration has been studied by Liao et al. (2005) that is presented in Figure 3. Here it is shown that the flammability region is slightly extended with the increase of ethane content in natural gas. LFL is almost 5% in volume and UFL is about 15%. The flammability limits are 3.0 to 12.5 in volume for ethane-air mixture. Their equivalent ratios are 0.512 and 2.506. The ratios are 0.486 and 1.707 with methane respectively. It is noted that the increase of ethane content in natural gas is extending the UFL in equivalence ratio but there is no remarkable change in LFL. Liao et al. (2005) show the effect of diluent ratio (ϕ_r) on flammability ratio. According to them, the increase of diluent ratio decreases the flammability region. The reason has been identified that the addition of diluents decreases the temperature of flame, which decreases the burning velocity. So, flammability limit goes narrower. Normally, CO_2 is more influential than N_2 addition. Shebeko et al. (2002) presented an analytical evaluation of flammability limits on ternary gaseous mixtures of fuel-air diluent. His prediction is shown in Figure 3 with dashed line.

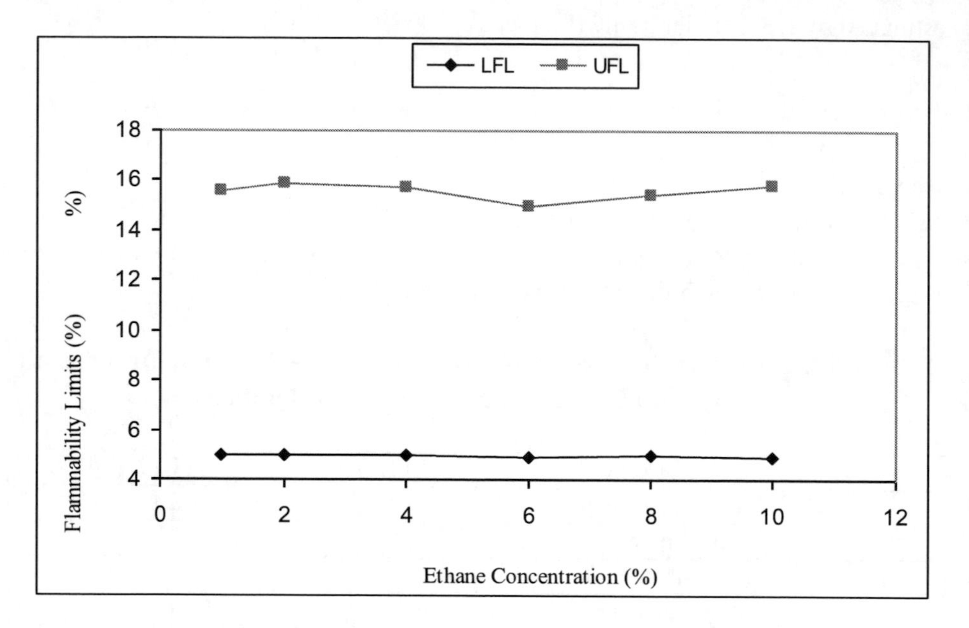

Figure 3. Dependence of NG flammability limits on ethane.

5. Individual Risk Based on Flammability

Figure 4 shows the incidental zone founded on basic fluid dynamics. The accidental scenario represents this incidental zone. If an explosion takes place for any reason, the incidental zone will be definitely covered by projectile theory of fluid dynamics. This concept is the basic difference from the model of Jo and Ahn (2005), which is shown in Figure 2. An accident due to flammability is considered here as the main cause of the incident. In Figure 4, OB is the maximum distance covered by the fire flame within which a fatality or injury can take place. BA and BC are the maximum distances traveled by the flame.

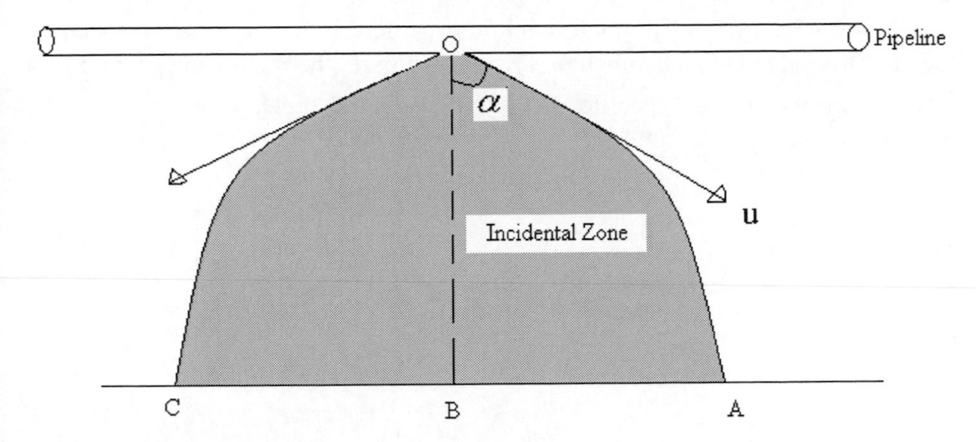

Figure 4. The relation of variables related with IRf .

The velocity of the natural gas evolved through the hole can be written as:

$$u = 1.273 \frac{q_{\min}}{d_{hole}^{2}} \quad (5)$$

where

$q_{\min}$ = Minimum gas flow rate evolved through the hole that causes an explosion $= f(u, d_{hole})$, $ft^3/\sec$

d_{hole} = Diameter of the hole through which gas passes, *ft*

Hazard distance or maximum distance covered by gas particles can be written as:

$$h_{\max} = \frac{1}{2} ut \cos\alpha \quad (6)$$

where:

$h_{\max}$ = Hazard distance, *ft*

u = Velocity of gas, $ft/\sec$

t = Travel time to reach the hazard distance, *sec*

α = Angle between velocity of gas and hazard distance, *degree*

Figure 4 shows the geometric relations among the variables in specified location from a natural gas pipeline. From this figure, the interacting section of a straight pipeline, $l_{\pm}$ from specified location, B, and the angle, α are estimated by the following equations:

$$l_{\pm} = \frac{1}{2} ut \sin \alpha \tag{7}$$

and

$$\alpha = \tan^{-1}\left(\frac{l}{h_{\max}}\right) \tag{8}$$

The individual risk (IR_f) due to flammability limit in a natural pipeline can be written as:

$$IR_f = \sum_i \frac{\varphi_i}{100} \int_{-l}^{+l} \int_0^{h_{\max}} (UFL_i - LFL_i) dh dl \tag{9}$$

where

φ_i = The failure rate per unit length of the pipeline associated with the accident scenario *i* due to flammability

l = Pipeline length, *ft*

UFL, LFL = Upper and lower flammability limit

$l_{\pm}$ = Ends of the interacting section of the pipeline in which an accident poses hazard to the specified location, *ft*

Figure 5 shows the number of incidents with pipeline distance from the source of gas. The data has been collected from the US office of pipeline safety, incident summary statistics from 1986 to August, 2005 (Web site 1). In this figure, the number of incidents are oscillating pattern within the region of 67775 and 259136 miles, however, beyond this distance, the rate of incidents show an abnormal pattern. It might be the cause of other factors, such as natural disaster, human activities.

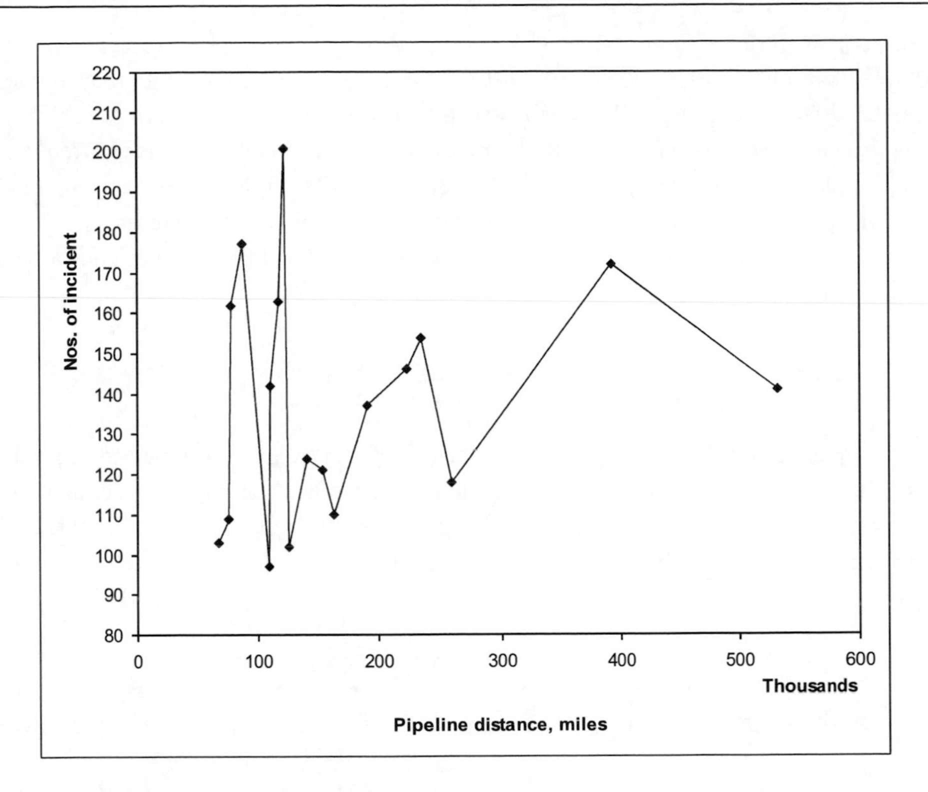

Figure 5. Incident related with pipeline distance.

There is no available model that handles both flammability limit and lethality for measuring individual risk. It is difficult to get the data for the particular reason of accidental scenario due to flammability. Based on available information and data dealing with this issue, the proposed model can be easily verified with any sets of data with confidence. In this study, 10% of accidental scenarios are assumed to be due to flammability (Web site 1). Using these data, the proposed model (Equation 9) is tested and results are shown in Figure 6. It shows the individual risk due to flammability with number of injuries. The normal trend of the curve is increasing with the increase of number of incident which leads to a separate scenario of accidents due to flammability. This chart also shows that there is a great impact of flammability on accidental scenario.

Figure 7 shows the probability of individual risk due to flammability with pipeline distance using Equation 9. Here it has been assumed that the *UFL* and *LFL* are 15.6 and 5.0 for the calculation. $q_{\min}$ is considered as $1\,ft^3/\sec$, $\alpha = 45^0$, $t = 1$ min and $d_{hole} = 0.5$ ft for a case study. The available literature shows that the maximum value of h is 66 *ft* and l is 99 *ft* . Here the calculation shows that h is 80.5 ft and l is 129.93 ft. These values seem to be quite reasonable. The individual risk due to flammability is decreasing with pipeline distance from the gas supply center. However, the trend is quite unpredictable and more frequent in an accident scenario within the pipeline range of 124,931 miles. This graph also shows that there is a great impact of flammability on accidental scenario.

Now combining Equation 1 and 9, a combined individual risk in a natural gas pipeline is obtained:

$$IR_T = IR + IR_f \tag{10}$$

This equation represents a true scenario of an individual risk due to lethality and flammability of natural gas. The lethality of natural gas pipeline depends on operating pressure, pipeline diameter, distance from the gas supply to pipeline and the length of the pipeline from the gas supply or compressing station to the failure point.

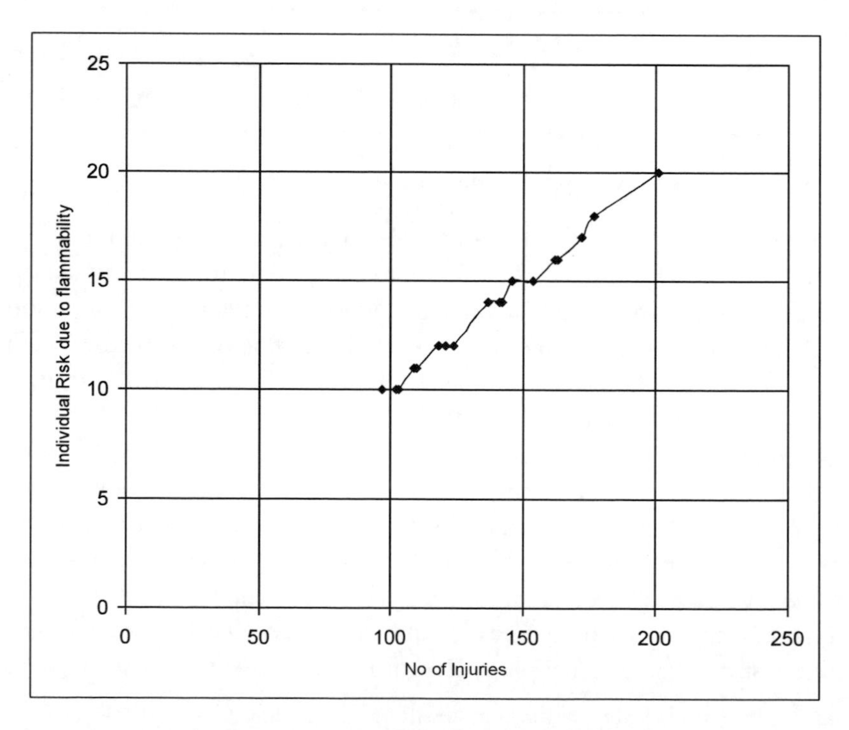

Figure 6. Individual risk due to flammability with number of injuries.

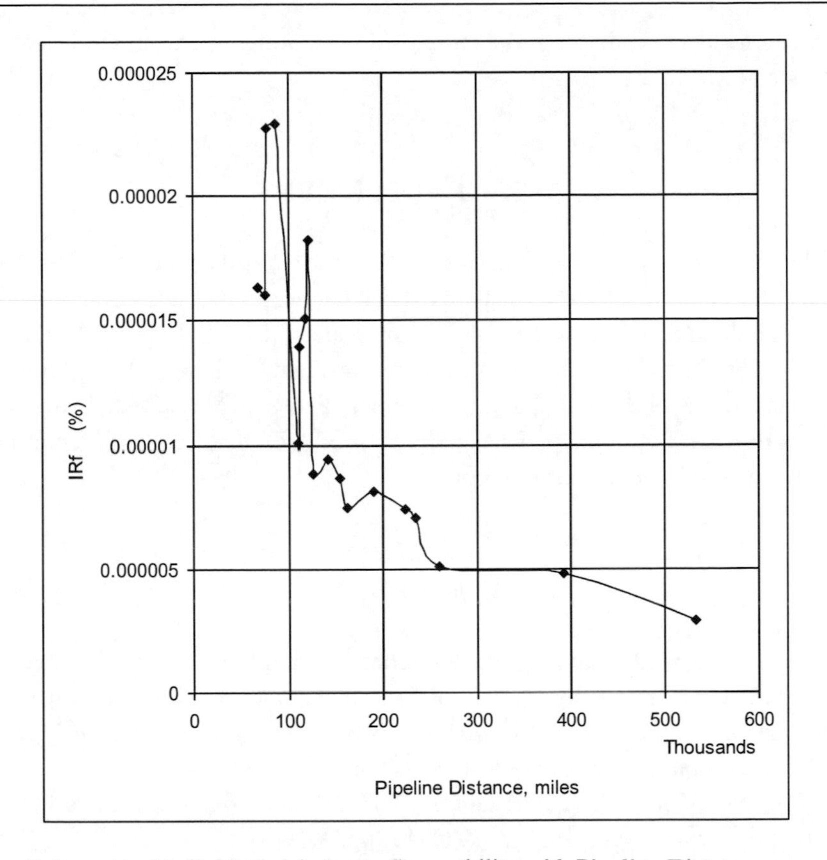

Figure 7. Percent of Individual risk due to flammability with Pipeline Distance.

CONCLUSION

Extensive pipeline network for natural gas supply system possesses many risks. Appropriate risk management should be followed to ensure safe natural gas pipelines. Individual risk is one of the important elements for quantitative risk assessment. Considering the limitations in conventional risk assessment, a novel method is developed for measuring individual risk combining all probable scenarios and parameters associated with practical situations taking into account gas flammability. These parameters can be calculated directly by using the pipeline geographical and historical data. By using the proposed method, the risk management can be more appealing from practical point of view. The proposed model is found to be innovative using pipeline and incident statistical data. The method can be applied to pipeline management during the planning, design, and construction stages. It may also be employed for maintenance and modification of a pipeline network.

ACKNOWLEDGMENT

The authors would like to thank the Atlantic Canada Opportunities Agency (ACOA) for funding this project under the Atlantic Innovation Fund (AIF). The authors would like to thank Dr. M. Zubair Kalam, SCAL Expert, ADCO, Abu Dhabi, UAE; Dr. Ali S. Al-Bemani, Vice-President, Sultan Qaboos University, Oman; and Dr. Mohammad Tamim, Professor, PMRE, Bangladesh University of Engineering and Technology, Dhaka, Bangladesh for their valuable contribution while accomplishing this research project.

REFERENCES

Arnaldos, J., Casal, J., Montiel, H., Sa´nchez-Carricondo, M., and Vı´lchez, J.A. (1998), 'Design of a computer tool for the evaluation of the consequences of accidental natural gas releases in distribution pipes', *Journal of Loss Prevention in the Process Industries*, 11: 135–148

Jo, Y.-D. and Ahn, B.J. (2002) '*Analysis of hazard area associated with high pressure natural-gas pipeline',* Vol. 15, pp. 179.

Jo, Y-D and Ahn, B. J. (2005) 'A method of quantitative risk assessment for transmission pipeline carrying natural gas', *Journal of Hazardous Materials*, A123, pp. 1–12.

John, M., Chris, B., Andrew, P. and Charlotte, T. (2001), 'An Assessment of Measures in Use for Gas Pipeline to Mitigate against Damage Caused by Third Party Activity' Printed and Published by the health and safety executive, C1 10/01

Kenneth, L. Cashdollar, Isaac, A. Zlochower, Gregory, M. Green, Richard, Thomas, A. and Hertzberg, M. (2000), 'Flammability of methane, propane, and hydrogen gases', *J. Loss Prev. Process. Ind.* 13, pp. 327–340.

Kevin, J. L, Isaac, A. Zlochower, Kenneth, L. Casdollar, Sinisa, M. Djordjevic, and Cindy, A. Loehr (2000), 'Flammability of gas mixtures containing volatile organic compounds and hydrogen', *J. Loss Prev. Process. Ind.* 13, pp. 377–384.

Khan, M.I., Lakhal, Y.S., Satish, M., and Islam, M.R., 2006. 'Towards Achieving Sustainability: Application of Green Supply Chain Model in Offshore Oil and Gas Operations', *Int. J. Risk Assessment and Management*: submitted.

Liao, S.Y., Cheng, Q., Jiang, D.M. and Gao, J. (2005) 'Experimental study of flammability limits of natural gas–air mixture', *Journal of Hazardous Materials*, Vol. B119, pp. 81–84.

Mishra, D.P., and Rahman, A. (2003), 'An experimental study of flammability limits of LPG/air mixtures', *Fuel* 82, pp. 863–866.

Pfahi, U.J., Ross, M.C., and Shepherd, J.E. (2000), 'Flammability limits, Ignition energy, and flame speeds in H_2–CH_4–NH_3–N_2O–O_2– N_2 mixtures, Combust', *Flame* 123, pp. 140–158.

Ramanathan, R. (2001) 'Comparative risk assessment of energy supply technologies: a data envelopment analysis approach', *Energy*, Vol. 26, pp.197–203.

Shebeko, Y.N., Fan, W., Bolodian, I.A., and Navzenya, V. Yu. (2002), 'An analytical evaluation of flammability limits of gaseous mixtures of combustible-oxidizer-diluent', *Fire Safety Journal,* 37 pp. 549-568.

Takahashi, A., Urano, Y., Tokuhashi, K., and Kondo, S. (2003), 'Effect of vessel size and shape on experimental flammability limits of gases, *J. Hazard. Mater',* A105, pp. 27–37.

TNO Purple Book, (1999), '*Guideline for Quantitative Risk Assessment, Committee for the Prevention of Disasters'*, The Netherlands, (Chapter 6).

Vanderstraeten, B., Tuerlinckx, D., Berghmans, J., Vliegen, S., Oost, E. Vant, and Smit, B. (1997), 'Experimental study of the pressure and temperature dependence on the upper flammability limit of methane/air mixtures', *J. Hazard. Mater.* 56, pp. 237–246.

Web site 1: http://ops.dot.gov/library/saferep/saferep.htm accessed on October 13, 2005

Wierzba, I., and Ale, B.B. (2000), 'Rich flammability limits of fuel mixtures involving hydrogen at elevated temperatures', *Int. J. Hydrogen Energy* 25, pp. 75–80.

In: Natural Gas Systems
Editor: Rafiq Islam
ISBN: 978-1-61324-158-5

Chapter 5

HUMAN HEALTH RISKS ASSESSMENT DUE TO NATURAL GAS PIPELINES EXPLOSIONS

M. Enamul Hossain, M. Ibrahim Khan, Chefi Ketata and M. Rafiqul Islam*

Department of Civil and Resource Engineering
Dalhousie University, Halifax, NS B3J-2X4, Canada

ABSTRACT

Natural gas is transported mainly by pipelines throughout the world. Therefore it is necessary to assess and manage the resulting risks regarding human health issues due to gas toxicity and flammability. It is possible to assess the risk of irreversible damage to a human being for any accidental scenario by introducing specific vulnerability functions. Events such as flash fire, vapor cloud explosion, and fire can be understood by the maximum predicted amount of vapor within the flammability limits for the entire history of its dispersion. Another danger to human health lies in the flammability of natural gas transportation systems. A human health risk assessment study in the event of such an accident has been carried out in this paper. In this study, a 1 to 20% accidental rate is considered for assessing individual risk due to flammability. A newly developed flammability risk management model is used in the present study. The research shows that the individual risk due to the flammability of natural gas is not more than an 18

* Corresponding author: Email: rafiqul.islam@dal.ca

percent human health hazard. The findings of this study will be helpful to improve health hazard risk management and remediation.

Keywords: Human health; natural gas pipelines; flammability limit; individual risk.

1. INTRODUCTION

Risk assessment addresses pipeline safety, environmental protection, financial management, project or product development, and many other areas of business performance. In this case, risk assessment considers pipeline safety in relation to protecting human life, the environment and property due to pipeline failure accidents. A pipeline can fail and release oil or natural gas into the environment and may cause many problems including environmental degradation, and loss of human life due to flammability and damaging effects of pollution as well.

The goal of risk assessment is to assess the likelihood that a possible threat could lead to a failure at a particular location on the pipeline and what the consequences might be. This assessment is conducted by identifying the specific characteristics of the pipeline at any given location, along with the unique characteristics of the area around the pipeline. The susceptibility of the pipeline to failure and its impacts is dependent on numerous characteristics, such as the type and condition of the pipe's coating, condition of the soil around the pipe, distance of pipeline from locality, and the contents of pipeline. For instance, water content of gas in pipes usually is one of the biggest reasons of corrosion in presence of other active components (Knickerbocker, 2006).

To determine the individual risk of an explosion hazard, flammability limits data are essential in a natural gas pipeline. Flammability limits are commonly used indices to represent the flammability characteristics of gases. The flammability limit criterion, and other related parameters have been broadly discussed in the available literature (Vanderstraeten et al., 1997; Kenneth et al., 2000; Kevin et al., 2000; Pfahi et al., 2000; Wierzba and Ale, 2000; Mishra and Rahman, 2003; Takahashi et al., 2003; Liao et al., 2005a; Liao et al., 2005b).

Hossain et al. (2008) studied the flammability and individual risk assessment for natural gas pipelines. They developed a comprehensive model for the individual risk assessment where the flammability limit with existing individual risk for an accidental scenario has been combined. Their model applies to the major accidental area within a locality surrounded by pipelines, and for any natural gas pipeline risk assessment scenario. Hossain et al. (2008) also verified the model using available field data. However, they assume a 10% accident occurrence due to flammability in a natural gas pipeline accident. The accidental scenario may be any percentage within a limited value. The present study applies the same model to verify different accidental scenarios. For a case study, 1%~20% accidental rates are considered in this paper, a conservative figure in risk assessment.

In the case of risk assessment, Fabbrocino et al. (2005) reported that the assessment must be as conservative as possible. They also added that whatever the final assessment: "worst case" should always be considered. When uncertainties are faced, the deterministic assessment even in the framework of probabilistic safety assessment should be taken into account. This approach is particularly effective, when late or early ignition assumption is considered in risk assessment (Fabbrocino et al., 2005).

The human health risk assessments determine how threatening a pipeline accident will be to human health. The main objective of human health risk assessment is to determine a safe level of contaminants or releases of toxic compounds, such as oil and natural gas from a pipeline. In the case of individual humans, there is a standard at which ill health effects are unlikely. It also estimates current and possible future risks. This paper examines the individual risk of natural gas flammability on human health. The goal of this study is to manage risks to acceptable levels, and recommend a method for risk managers to incorporate risk assessment information for the planning and developing of pipeline networks.

2. Risk Management

Pipeline risk management deals with pipeline system failures due to:

1) Corrosion.
2) Cracking.
3) Material degradation or defects.
4) Third-party damage such as sabotage.
5) Earth movements.

It is paramount to assess and manage pipeline risks by considering the potential consequences of pipeline failures. The possible potential consequences are:

1) Damage to human health and safety including injuries and fatalities.
2) Property damage.
3) Environmental damage.

Long-term exposure to hazardous material is the paramount risk regarding long-term damage to human health such as asthma and cancer. Safety risk is the acute risk related to short-term damage to the human body such as burns, injuries, and death due to an accident or exposure to explosion.

Risk management is the process that examines the following phases (see Figure 1):

1) Identification.
2) Assessment.
3) Remediation.
4) Evaluation.
5) Maintenance.

Risk identification deals with:

1) Site location.
2) Hazard identification.
3) Risk analysis.

Risk assessment involves estimating various health and safety risk parameters such as the individual risk. There are two types of risk assessment:

1) Qualitative.
2) Quantitative.

The risk remediation stage addresses the following steps:

1) Strategy proposal.
2) Strategy implementation.

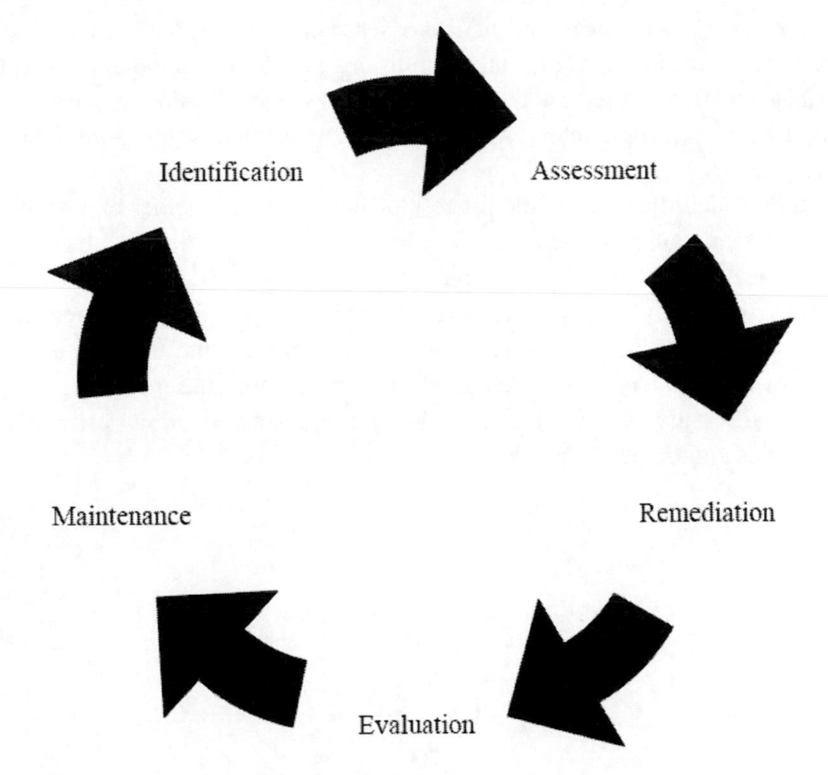

Figure 1. Risk management for natural gas pipelines.

3. Human Health Risk Assessment

The components of human health risk assessments are: planning and scoping, exposure assessment, acute hazards, toxicity and risk characterization. The main components of human health risk assessment are shown in Figure 2. There are four different steps in assessing human health risk, which are Planning and Scoping, Exposure Assessment, Acute Hazard Assessment, and Risk Characterization. For efficient risk assessments the 'planning and scoping' of the information and data are needed. It should be done before the field investigations and site characterization.

The second step of human health risk assessment is 'exposure assessment' (see Figure 2) that is the contact of natural gas to the human. This process considers how much time, duration and frequencies of the chemical contact with a human in the past, present and future. The 'exposure assessment' step should be done following step one. This step should be conducted just once, but if necessary it can be repeated for accuracy of the assessment. “In the case of human risk assessment, ‘acute hazards’ mean the conditions that create the potential for injury or damage to occur due to an instantaneous or short duration exposure to the effects of an accidental release. In this study, it is mainly the flammability of natural gas.

‘Hazard identification’ is the process of determining whether exposure to the natural gas can cause an increase in the incidence of a particular adverse health effect. Generally, it is done by the dose responses of particular chemicals. However, this study considered the flammability. The ‘Risk Characterization’ process is the synthesis of results of all other steps and the determination how dangerous the accident is to pipelines. It also considered the major assumptions, and scientific judgments. Finally, the risk characterization estimates the uncertainties embodied in the assessment.

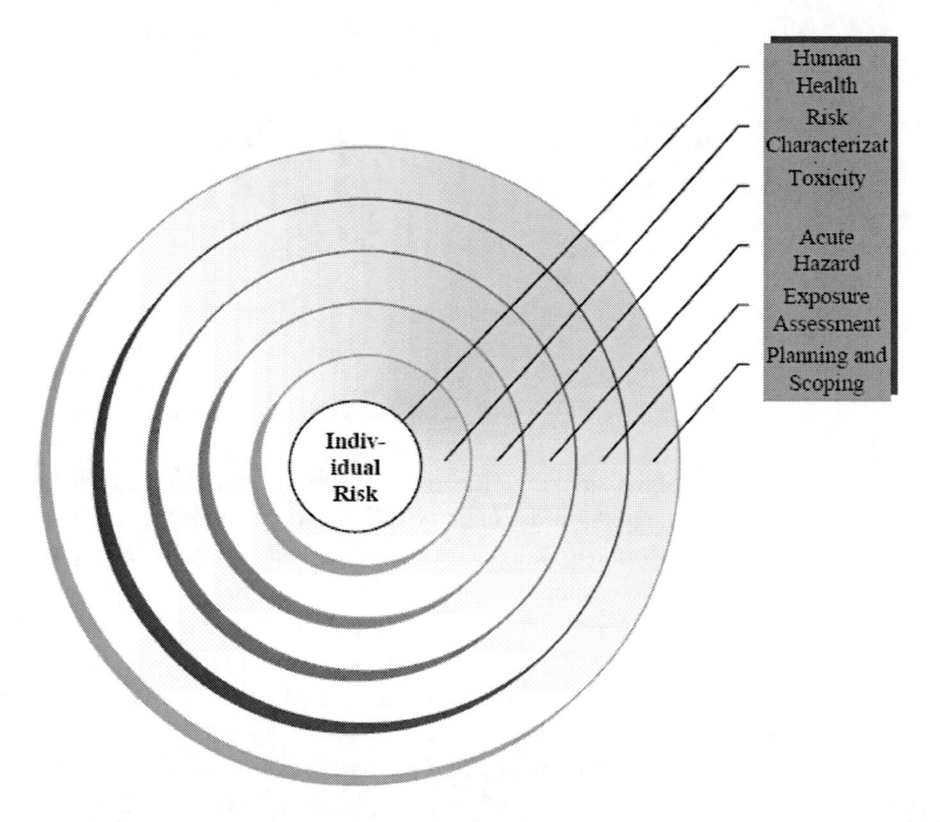

Figure 2. Different components of human health risks assessment.

3.1. Human Health Risk Levels

In the risk assessment, human health is the major concerning issue, but there are other factors to consider as well, such as ecological risk assessment. Pipelines carry natural gas which has numerous toxic compounds that might directly and indirectly cause risks to human health. Pipelines carry natural gas that contains methane, ethane, propane, iso – butane, normal – butane, iso – pentane, normal – pentane, hexanes plus, nitrogen, carbon dioxide, oxygen, hydrogen, hydrogen sulfide. Sour gas contains larger amount of hydrogen sulfide. In the case of any pipeline accident all of these compounds are released. Due to flammability and exposure of all of these compounds, different levels of risk can take place. Very recently (May 2006), more than 150 people were killed due to flammability in case of pipelines. It is reported that a ruptured fuel pipeline exploded and caught fire near Nigeria's largest city, Lagos (IRIN, 2006). This pipeline transports fuel from a depot at the Lagos port for domestic use inland. Victims were inhabitants of poor fishing villages. Pipeline accidents are common in third world counties,

Table 1. Human health risk levels

Risk levels	Concentration (ppm)	Effects
Negligible or no- Risk	0.01-0.3	Odor threshold (highly variable)
Minimal Risk	1-5	Moderate offensive odor, may be associated with nausea, tearing of the eyes, headaches or loss of sleep with prolonged exposure; healthy young male subjects experience no decline in maximal physical work capacity
Slightly Moderate Risk	10 - 8 h	Occupational exposure limit
Moderate Risk	20-50	Ceiling occupational exposure limit and community evacuation level, odor very strong
Risk	100	Eye and lung irritation; olfactory paralysis, odor disappears
High Risk	150-200	Sense of smell paralyzed; severe eye and lung irritation
Severe Risk	250-500	Pulmonary edema may occur, especially if prolonged
Extremely High	500	Serious damage to eyes within 30 min; severe lung irritation; unconsciousness and death within 4-8 h; amnesia for period of exposure; "knockdown"
Critical Level	1000	Breathing may stop within one or two breaths; immediate collapse

Source: Guidotti, 1994.

such as Nigeria, oil rich African nation. In 1998, it is reported that more than 1000 people died due to a flammability accident in Jesse, near the oil town of Warri, Niger Delta (IRIN, 2006).

In above accident report, it is revealed that due to strong flammability the fate is certainly death, but exposures to other components, such as hydrogen sulfide have different risk levels. In Table 1, different risk levels cause by the hydrogen sulfide is shown. This phenomenon needs to be considered seriously in case of sour gas, where hydrogen sulfide concentration is higher. Generally, the typical sulphur content is 5.5 mg/m^3, which includes the 4.9 mg/m^3 of sulphur in the odorant (mercaptan) added to gas for safety reasons.

3.2. Combustion Properties of Natural Gas

As mentioned earlier, natural gas has an extreme risk of flammability due to its composition. To understand the flammability risk of natural gas, the combustion properties of natural gas are presented in Table 2 (Data source: Union Gas, 2006). It is noted that the combustion properties of gas depends on its compositions, but a general estimations is shown in Table 2. The properties shown are an overall average on the Union Gas system (Union Gas, 2006).

Table 2. Typical Combustion properties of Natural Gas

Ignition Point	593 ºC
Flammability Limits	4% - 16% (vol. % in air)
Theoretical Flame Temperature (stoichiometric air/fuel ratio)	1960 ºC (3562 ºF)
Maximum Flame Velocity	0.3 m/s
Relative density (specific gravity)	0.585

4. Individual Risk Based on Flammability

Hossain et al. (2008) have shown the concept of individual risk due to flammability at a locality where dense populations live in. Figure 3 has been redrawn from this reference where detailed analysis has been presented. An accident due to flammability is considered here as the main cause of the incident. In Figure 3, OB is the maximum distance covered by the fire flame within which a fatality or injury can take place. BA and BC are the maximum distances traveled by the flame.

The individual risk (IR_f) due to flammability limit in a natural pipeline can be written as:

$$IR_f = \sum_i \frac{\varphi_i}{100} \int_{-l}^{+l} \int_0^{h_{max}} (UFL_i - LFL_i)\, dh\, dl \tag{1}$$

and the total individual risk can be written as;

$$IR_T = IR + IR_f \tag{2}$$

where,

φ_i = The failure rate per unit length of the pipeline associated with the accident scenario i due to flammability

l = Pipeline length, *ft*

UFL, LFL = Upper and lower flammability limit

$l_\pm$ = Ends of the interacting section of the pipeline in which an accident poses hazard to the specified location, *ft*.

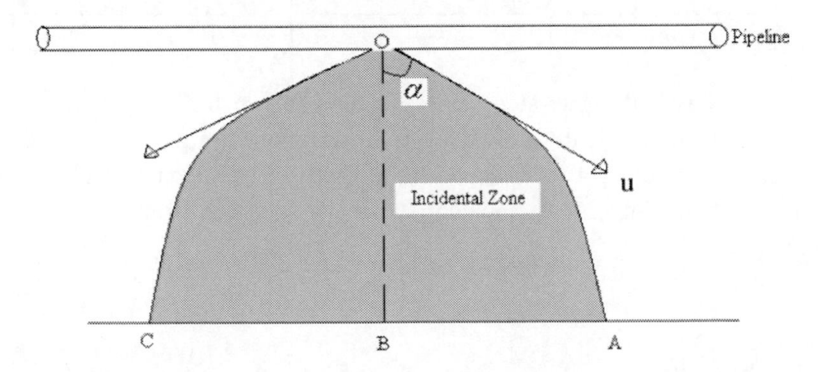

Figure 3. The relation of variables related with IRf (redrawn from Hossain et al., 2008).

Table 3 shows the different data for number of fatalities/injuries and number of fatalities/injuries due to natural gas flammability accident in pipeline from 1985 to 2005. The data has been collected from the department of pipeline safety of U.S.A.

Table 3. Number of injury and flammability data for different percentage (Data Source: Website 1)

Fatality/ Injury	Fatality/injury due to natural gas flammability									
	1%	3%	6%	8%	10%	12%	14%	16%	18%	20%
97	0.97	2.91	5.82	7.76	9.7	11.64	13.58	15.52	17.46	19.4
102	1.02	3.06	6.12	8.16	10.2	12.24	14.28	16.32	18.36	20.4
103	1.03	3.09	6.18	8.24	10.3	12.36	14.42	16.48	18.54	20.6
109	1.09	3.27	6.54	8.72	10.9	13.08	15.26	17.44	19.62	21.8
110	1.1	3.3	6.6	8.8	11	13.2	15.4	17.6	19.8	22
118	1.18	3.54	7.08	9.44	11.8	14.16	16.52	18.88	21.24	23.6
121	1.21	3.63	7.26	9.68	12.1	14.52	16.94	19.36	21.78	24.2
124	1.24	3.72	7.44	9.92	12.4	14.88	17.36	19.84	22.32	24.8
137	1.37	4.11	8.22	10.96	13.7	16.44	19.18	21.92	24.66	27.4
141	1.41	4.23	8.46	11.28	14.1	16.92	19.74	22.56	25.38	28.2
142	1.42	4.26	8.52	11.36	14.2	17.04	19.88	22.72	25.56	28.4
146	1.46	4.38	8.76	11.68	14.6	17.52	20.44	23.36	26.28	29.2
154	1.54	4.62	9.24	12.32	15.4	18.48	21.56	24.64	27.72	30.8
162	1.62	4.86	9.72	12.96	16.2	19.44	22.68	25.92	29.16	32.4
163	1.63	4.89	9.78	13.04	16.3	19.56	22.82	26.08	29.34	32.6
172	1.72	5.16	10.32	13.76	17.2	20.64	24.08	27.52	30.96	34.4
177	1.77	5.31	10.62	14.16	17.7	21.24	24.78	28.32	31.86	35.4
201	2.01	6.03	12.06	16.08	20.1	24.12	28.14	32.16	36.18	40.2

Figure 4 has been generated using the data shown in Table 3. It shows the number of incidents with individual risk due to flammability for different percentage of flammability risk at pipeline. The data has been collected from the U.S. office of pipeline safety, incident summary statistics from 1986 to August, 2005 (Web site 1).

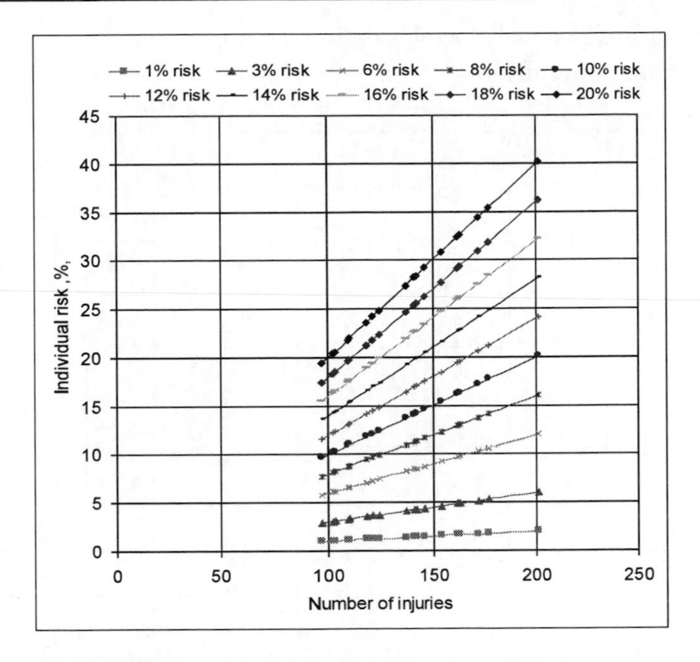

Figure 4. Individual risk due to flammability with number of injuries.

In this figure, the individual risk is increasing to a steeper trend when human health hazard risk due to flammability injuries are increased. It means that the individual risk factor is very much influenced by the flammability risk factor within the contour locality.

At present, there are many models available to investigate individual risk (John et al., 2001; Jo et al., 2002 and 2005; Fabbrocino et al., 2005). However, there is no model available that handles both flammability limit and lethality for measuring individual risk for human health hazard. It is difficult to get data for the accidental scenario due to flammability. Based on available information and data dealing with this issue, the Hossain et al., (2008) model can be easily used to verify with any sets of data with confidence. In this study, 1~20% of accidental scenarios are considered to be due to flammability (web site 1). Using these data, the model (Equation 1) is tested and results are shown in Figures 5 and 6. Here it has been assumed that the *UFL* and *LFL* are 15.6 and 5.0 for the calculation. $q_{\min}$ is considered as $1\,ft^3/\sec$, $\alpha = 45°$, $t = 1$ min and $d_{hole} = 0.5$ ft for a case study. Jo and Ahn (2002) showed that the maximum value of *h* was 66 *ft* and *l* was 99 *ft*. They used the triangular explosion concepts. Here the calculation shows that *h* is 80.5 ft and *l* is 129.93 ft (Hossain et al., 2008). These values seem to be quite reasonable since the projectile explosion model proposed in the previous paper is more precise and convenient.

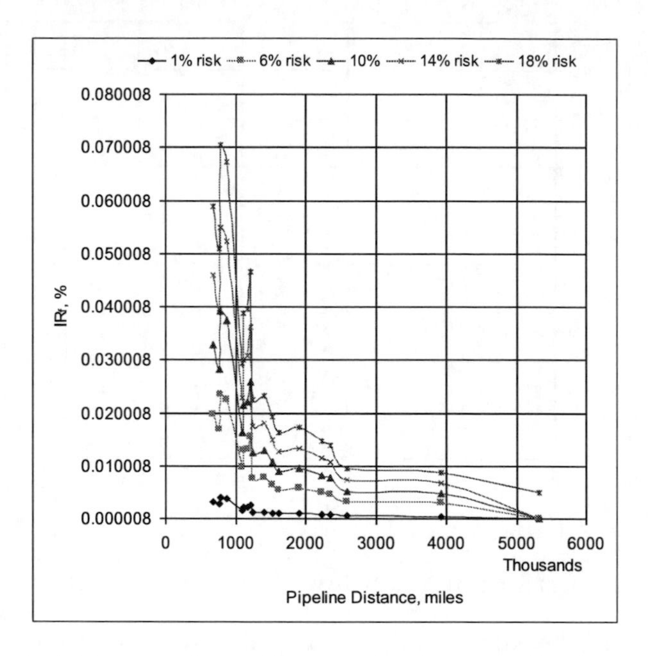

Figure 5. Individual risk due to flammability as a function of pipeline distance.

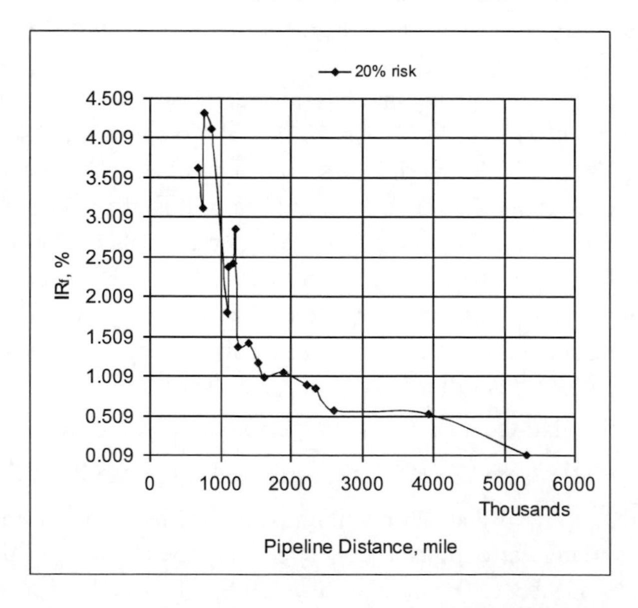

Figure 6. Individual risk due to flammability as a function of pipeline distance.

Table 4 shows the different individual risk due to flammability data for different pipeline distances. The flammability data has been calculated using equation (1). The pipeline data that causes the fatalities/injuries to natural gas in pipeline accidents are from 1985 to 2005. The data has been collected from the department of pipeline safety of USA.

Table 4. Individual risk due to flammability with pipeline distance

Pipeline distance, (miles)	Individual risk due to flammability					
	1.0%	6.0%	10.0%	14.0%	18.0%	20.0%
5320616	8.72E-06	2.33E-05	3.49E-05	4.65E-05	0.005078	0.009638
3928390	0.00048742	0.002927	0.004879	0.00683	0.008781	0.538715
2591365	0.0005272	0.003163	0.005272	0.00738	0.009489	0.582692
2339883	0.00077257	0.004634	0.007723	0.010812	0.013901	0.853882
2229440	0.00081789	0.004908	0.00818	0.011452	0.014724	0.903967
1905511	0.00095825	0.005749	0.009582	0.013415	0.017248	1.059106
1625284	0.0009052	0.005431	0.009051	0.012672	0.016292	1.000471
1534665	0.00107209	0.00643	0.010716	0.015003	0.01929	1.184929
1407148	0.00129314	0.007748	0.012913	0.018078	0.023243	1.429245
1249316	0.00124893	0.007484	0.012474	0.017464	0.022453	1.380382
1213143	0.00258628	0.015525	0.025874	0.036224	0.046574	2.858489
1173612	0.00219945	0.0132	0.021999	0.030799	0.039599	2.430938
1107880	0.00215524	0.012905	0.021509	0.030112	0.038716	2.382075
1095067	0.00163577	0.00981	0.016351	0.022891	0.029431	1.807933
867581	0.00373574	0.022426	0.037377	0.052328	0.067279	4.128928
776574	0.00391258	0.0235	0.039167	0.054834	0.070501	4.324381
759404	0.00282944	0.017002	0.028337	0.039672	0.051006	3.127236
677750	0.00328259	0.019667	0.032778	0.04589	0.059001	3.628082

It shows the individual risk due to flammability with pipeline distance. The normal trend of the curve decreases with the increase of pipeline distance which leads to a separate scenario of accidents due to flammability. This chart also shows the impact of flammability on an accidental scenario. The interesting outcome of this model shows that human health hazard risk due to flammability in individual risk assessment of natural gas is limited by 18% of the total risk factor (see Figures 5 and 6). These figures have been generated using the data shown in Table 4. Beyond 18% of total individual risk, the figures do not fit with the other percentages of risk and the values of these calculations are not realistic (see

Figure 6). This information simply means that the human health hazard individual risk due to flammability of natural gas does not go beyond 18% of individual risk.

Conclusions

Extensive pipeline networks for natural gas supply systems possess many risks. Appropriate risk management should be followed to ensure safe natural gas pipelines. Individual risk is one of the important elements for quantitative risk assessment. Considering the limitations in conventional risk assessment, a novel method is developed for measuring individual risk combining all probable scenarios and parameters associated with practical situations taking into account gas flammability. These parameters can be calculated directly by using the pipeline geographical and historical data. By using the proposed method, the risk management can be more appealing from practical point of view. The proposed model is found to be innovative using pipeline and incident statistical data. The method can be applied to pipeline management during the planning, design, and construction stages. It may also be employed for maintenance and modification of a pipeline network.

Acknowledgment

The authors would like to thank the Atlantic Canada Opportunities Agency (ACOA) for funding this project under the Atlantic Innovation Fund (AIF).

References

Cashdollar, K.L., Zlochower, I.A., Green, G.M., Thomas, R.A. and Hertzberg, M. (2000) 'Flammability of methane, propane, and hydrogen gases', *J. Loss Prev. Process. Ind.*, Vol. 13, pp. 327-340.

Fabbrocino, G., Iervolino, I., Orlando, F. and Salzano, E. (2005) 'Quantitative risk analysis of oil storage facilities in seismic areas', *Journal of Hazardous Materials*, Vol. 123, Issues 1-3, pp. 61–69.

Guidotti, T.L. (1994) 'Occupational exposure to hydrogen sulfide in the sour gas industry: Some unresolved issues', *Int Arch Occup Environ Health*, Vol. 66, pp. 153-160.

Hossain, M.E., Ketata, C., Khan M.I. and Islam M.R. (2008) 'Flammability and individual risk assessment for natural gas pipelines', *Advances in Sustainable Petroleum Engineering Science*, 1(1), article in press.

IRIN (2006) *NIGERIA: More than 150 killed in pipeline blast*, UN Office for the Coordination of humanitarian Affairs, www.IRINnews.org.

Jo, Y.-D. and Ahn, B.J. (2002) 'Analysis of hazard area associated with high pressure natural-gas pipeline', Vol. 15, pp. 179.

Jo, Y-D and Ahn, B. J. (2005) 'A method of quantitative risk assessment for transmission pipeline carrying natural gas', *Journal of Hazardous Materials*, Vol. 123, pp. 1–12.

John, M., Chris, B., Andrew, P. and Charlotte, T. (2001) *An assessment of measures in use for gas pipeline to mitigate against damage caused by third party activity*', Printed and published by the health and safety executive, C1 10/01.

Knickerbocker, B. (2006) 'Leak is latest of Alaska's pipeline woes', *The Christian Science Monitor*, August 9.

Liao, S.Y., Cheng, Q., Jiang, D.M. and Gao, J. (2005a) 'Experimental study of flammability limits of natural gas–air mixture', *Journal of Hazardous Materials*, Volume 119, Issues 1-3, pp. 81-84.

Liao, S.Y., Jiang, D.M., Huang, Z.H., Cheng, Q., Gao, J. and Hu, Y. (2005b) 'Approximation of flammability region for natural gas–air–diluent mixture', *Journal of Hazardous Materials*, Volume 125, Issues 1-3, pp. 23-28.

Liekhus, K.J., Zlochower, I.A., Cashdollar, K.L., Djordjevic, S.M. and Loehr, C.A. (2000) 'Flammability of gas mixtures containing volatile organic compounds and hydrogen', *Journal of Loss Prevention in the Process Industries*, Vol. 13, Issues 3-5, pp. 377-384.

Mishra, D.P. and Rahman, A. (2003) 'An experimental study of flammability limits of LPG/air mixtures', *Fuel*, Vol. 82, pp. 863–866.

Pfahi, U.J., Ross, M.C. and Shepherd, J.E. (2000) 'Flammability limits, ignition energy, and flame speeds in H_2–CH_4–NH_3–N_2O–O_2–N_2 mixtures', *Combustion and Flame*, Vol. 123, Issues 1-2, pp. 140-158.

Takahashi, A., Urano, Y., Tokuhashi, K. and Kondo, S. (2003), 'Effect of vessel size and shape on experimental flammability limits of gases', *Journal of Hazardous Materials,* Vol. 105, Issues 1-3, pp. 27-37.

Union Gas (2006) *Chemical Composition of Natural Gas*, Union Gas, Chatham, Ontario, Canada.

Vanderstraeten, B., Tuerlinckx, D., Berghmans, J., Vliegen, S., Van't Oost, E. and Smit, B. (1997) 'Experimental study of the pressure and temperature dependence on the upper flammability limit of methane/air mixtures', *Journal of Hazardous Materials*, Vol. 56, Issue 3, pp. 237-246.

Website 1 (2005) *http://ops.dot.gov/library/saferep/saferep.htm*, accessed on October 13, 2005.

Wierzba, I. and Ale, B.B. (2000) 'Rich flammability limits of fuel mixtures involving hydrogen at elevated temperatures', *Int. J. Hydrogen Energy*, Vol. 25, pp. 75-80.

INDEX

E

F

G

H

I

W